Contents

TARGETING SCIENCE YEAR 3 © PASCAL PRESS ISBN: 9781925726527

Targeting Science Year 3
Copyright © 2024 Pascal Press

ISBN: 9781925726527

Published by Pascal Press
PO Box 250
Glebe NSW 2037
www.pascalpress.com.au
contact@pascalpress.com.au

This edition of *Skill Sharpeners: Science* is published by arrangement with Evan-Moor Corporation, USA.

For sale in Australia and New Zealand.

Cover Design: Janice Bowles
Authors: Guadalupe Lopez, Lisa Vitarisi Mathews
Publisher: Lynn Dickinson
Editor Australian edition: Stella Tarakson
Typesetter: Stacey Grainger
Illustrator: Paul Lennon, *www.dreamstime.com.au*
Cover design: Janice Bowles

Printed in China by 1010 International Ltd.

Introduction

Welcome to your Year 3 *Targeting Science* activity book! It is packed with interesting and exciting activities to help you understand and enjoy science at home or school.

Targeting Science has been written to support the Australian Primary Science Curriculum Version 9.0 and is divided between:

- Biological Sciences
- Earth & Space Sciences
- Physical Sciences
- Chemical Sciences

The Science Understanding and Inquiry Skills elements (including Science as a Human Endeavour) are embedded.

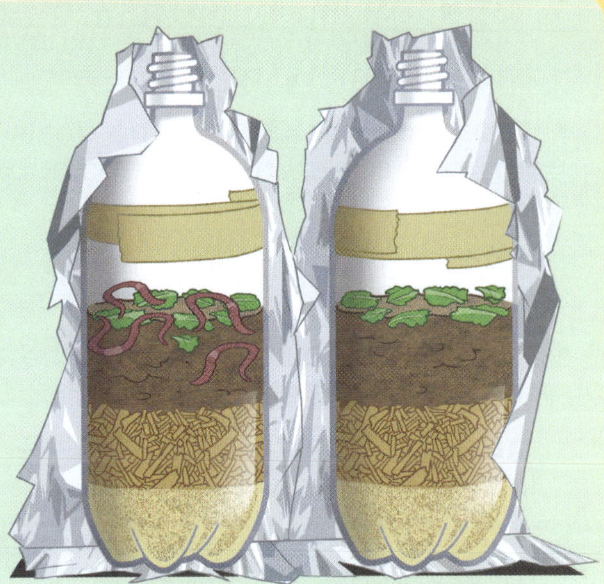

Hands-on activities provide opportunities to bring the science concepts to life. The exercises develop process skills such as observing, collecting, recording and organising information and making models of scientific events that happen in the natural world. All activities use easily sourced, inexpensive items.

Topics are introduced with explanations plus images to provide background information. Some lessons include a QR code where you can access a video for further explanations. You will be challenged to match, sort, label, sequence, analyse and answer questions with regular vocabulary practice puzzles throughout the book to help familiarise you with scientific terms. See the Glossary on the next page for some terms you may not already know. Answers are included at the end of the book.

FREE Teaching Guide.

This QR code links to a downloadable PDF of a Teaching Guide to support the material in this student workbook. The guide contains:

- Teaching plans and checklists.
- Graphic organisers.
- Material request forms (to send home for parents).
- Background information about each major topic as well as extension activities.

Use this QR code to access the FREE Teaching Guide.

Front cover - have you looked at the front cover? It depicts what life was like before a significant scientific discovery that led to the development of new technology. What is the difference between Science and Technology? The following quote sums it up nicely.

Science is the process of acquiring knowledge of natural phenomenon along with various reasons. Technology is the application of Science to the solution of problems. (Sismondo, 2018)

Glossary

- **energy** - energy is defined by what it does, where it comes from and how it behaves; it is invisible and responsible for all forces
- **geologist** - someone who deals with the earth's physical structure and substance, its history, and the processes that act on it
- **heat** - a form of energy called 'thermal'; it can cause one state of matter to change into another
- **horizons** - the line at which the earth's surface and the sky appear to meet
- **igneous** - rock that has solidified from volcanic lava or magma
- **insulated** - using materials which inhibit the transfer of heat, sound or electrical energy
- **matter** - a physical substance; anything that has mass and occupies space
- **metamorphic** - rock that has been transformed by heat, pressure or other natural forces
- **mineral** - a chemical compound (a combination of elelements) found naturally in the rocks of the Earth
- **molecule** - the smallest part of an element or compound, able to exist independently
- **non-living** - something that has never been alive
- **once-living** - something that was once a part of a living organism but is now dead
- **recycled** - waste materials that have been processed so that they can be reused
- **rock** - the solid material forming part of the Earth's surface exposed or underlying the soil
- **rock cycle** - a series of processes that outline how each of the major rock types – igneous, metapmorphic and sedimentaty – form and break down over time
- **sedimentary** - rocks which form when small pieces of rocks are worn off and then washed into low places where they gather in layers
- **soil** - soils result from the breakdown and weathering of organic material
- **thermometers** – used to measure temperature (how hot or cold a substance is

SAFETY
All of the investigations in this book are designed for kids to do safely in the home or classroom. However, adult supervision is recommended.

Year 3 Australian Science Curriculum Correlations

ACARA Code	Content description	Strand	Sub strand	Pages
AC9S3U01	Students learn to compare characteristics of living and non-living things and examine the differences between the life cycles of plants and animals	Science Understanding	Biological Sciences	2-29
AC9S3U02	Students learn to compare the observable properties of soils, rocks and minerals and investigate why they are important Earth resources	Science Understanding	Earth & Space Sciences	30-55
AC9S3U03	Students learn to identify sources of heat energy and examine how temperature changes when heat energy is transferred from one object to another	Science Understanding	Physical Sciences	56-67
AC9S3U04	Students learn to investigate the observable properties of solids and liquids and how adding or removing heat energy leads to a change of state	Science Understanding	Chemical Sciences	68-85
AC9S3H01	Students learn to examine how people use data to develop scientific explanations	Science as human endeavour	Nature and development of science	31, 56, 68
AC9S3H02	Students learn to consider how people use scientific explanations to meet a need or solve a problem	Science as human endeavour	Use and influence of science	2, 4. 5, 31, 56, 59, 65, 77, 78
AC9S3I01	Students learn to pose questions to explore observed patterns and relationships and make predictions based on observations	Science Inquiry	Questioning and Predicting	11, 36, 62, 66, 81
AC9S3I02	Students learn to use provided scaffolds to plan and conduct investigations to answer questions or test predictions, including identifying the elements of fair tests, and considering the safe use of materials and equipment	Science Inquiry	Planning and conducting	11, 36, 51, 52, 62, 66, 73
AC9S3I03	Students learn to follow procedures to make and record observations, including making formal measurements using familiar scaled instruments and using digital tools as appropriate	Science Inquiry	Planning and conducting	11, 36, 51, 52 62, 66, 74
AC9S3I04	Students learn to construct and use representations, including tables, simple column graphs and visual or physical models, to organise data and information, show simple relationships and identify patterns	Science Inquiry	Processing, modelling and analysing	30, 32, 60, 68, 75, 84
AC9S3I05	Students learn to compare findings with those of others, consider if investigations were fair, identify questions for further investigation and draw conclusions	Science Inquiry	Evaluating	22, 36, 80
AC9S3I06	Students learn to write and create texts to communicate findings and ideas for identified purposes and audiences, using scientific vocabulary and digital tools as appropriate	Science Inquiry	Communicating	3, 12, 17, 44, 52, 55, 58, 63, 64, 66, 80, 82, 83

Living, non-living or once living?

Find something near you and pick it up. Have a good look at it. Is it alive? Was it ever? How can you tell?

In science, there are rules that everyone follows to classify things. For example, scientists agreed that mammals all have warm blood, hair or fur and feed their young on milk.

Rules exist for classifying **living, non-living and once living** things.

Living things

Do:

- Respire (breathe)
- Take in nutrients (food)
- Respond to things around them (light, heat)
- Move by themselves
- Reproduce (have babies)
- Grow and change
- Remove waste

Non-living things

Do not:

- Respire (breathe)
- Take in nutrients (food)
- Respond to things around them (light, heat)
- Move by themselves
- Reproduce (have babies)
- Grow and change
- Remove waste

Once-living things

Used to:

- Respire (breathe)
- Take in nutrients (food)
- Respond to things around them (light, heat)
- Move by themselves
- Reproduce (have babies)
- Grow and change
- Remove waste

For something to be considered living, it has to do ALL of these things. For example, the flame from a candle moves by itself and needs oxygen to burn, but it does not reproduce, so it is non-living.
Water moves and changes from a solid to liquid and gas, but it does not breathe, eat or have babies. Water is non-living.
When you buy a bunch of flowers, pick a piece of fruit or eat at a wooden table, you are using something that used to be alive. It used to do all the things living things do, but now it doesn't. These things are once-living.

TARGETING SCIENCE YEAR 3 © PASCAL PRESS ISBN: 9781925726527

Can you make a list of the living, non-living and once-living things around you?

Can you put an 'L' in the box if it is living, 'N' if it is non-living and 'OL' for once-living?

Baby zebra

Water in a metal bucket

Cut flowers

Cheese on a board

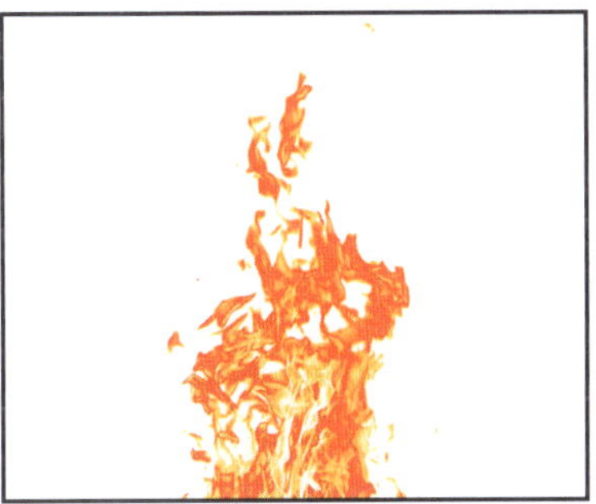

Fire

A bowl of fruit

A pot plant

A feather

A stack of rocks

Pick a living thing, a non-living thing and a once-living thing from the boxes above and explain why you grouped it the way you did:

1. Living thing: ____

   ____

2. Non-living thing: ____

   ____

3. Once-living thing: ____

   ____

Animals Grow and Change

Animal Group	Missing words	Complete the information
Mammals	• warm • milk • fur • mother	Mammals stay with their ______________ until they are able to look after themselves and no longer need her ______________. They have ______________ blood and hair or ______________.
Fish	• eggs • young • scales • blood	Fish have ______________ and cold ______________. Most do not look after their ______________. They can lay ______________ or have live babies.
Birds	• eggs • feathers • nest	Most birds lay their ______________ in a ______________. Almost all birds look after their young until they have ______________ and can look after themselves.
Reptiles	• scales • cold • crocodiles • live	All reptiles have ______________ and ______________ blood. Some reptiles lay eggs, and others have ______________ live babies. ______________ are one of the few reptiles that look after their young.
Insects	• three • stages • metamor-phosis	All insects have ______________ body parts. Most insects lay eggs. Many insects go through ______________ in their life cycles which involve change called ______________.
Amphibians	• eggs • skin • completely • adults	Amphibians have ______________ not scales. They start life as ______________ and ______________ change to become ______________.

TARGETING SCIENCE YEAR 3 © PASCAL PRESS ISBN: 9781925726527

Concept:

Animals grow and change.

Look at this mother cat and her kittens.

Cats are **mammals**. This means that mother cats grow their kittens inside their bodies. When the kittens are born, they are weak. The kittens drink milk from their mother's body. This is how it is with most mammals, including humans.

Look at this mother hen and her chick.

Life begins differently for **chicks**. This is because chickens are not mammals. A hen lays her eggs in a nest. The chicks break the eggs and **hatch**. They will grow into adult chickens, and the **life cycle** will continue.

TARGETING SCIENCE YEAR 3 © PASCAL PRESS ISBN: 9781925726527

Skill: Apply science vocabulary in context

Read the clue. Write the word.

1. Baby bird

___ ___ ___ ___ ___
(letter 2 = 1, letter 4 = 4)

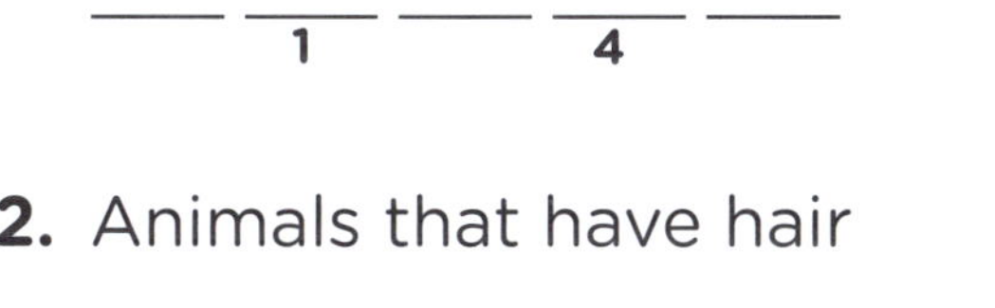

2. Animals that have hair or fur

___ ___ ___ ___ ___ ___ ___
(letter 2 = 2)

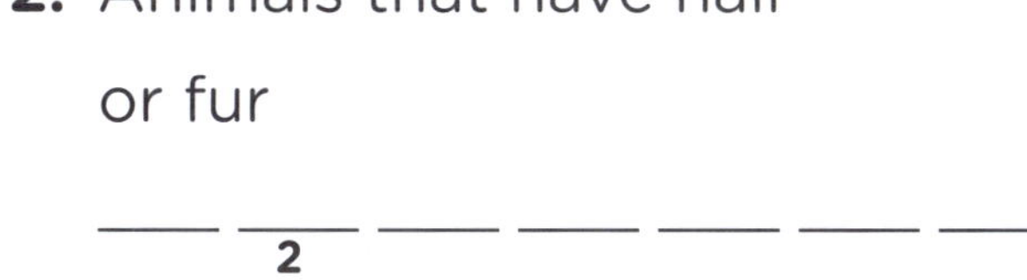

3. A baby mammal drinks this from its mother.

___ ___ ___ ___

4. A place where hens lay their eggs

___ ___ ___ ___
(letter 2 = 5, letter 3 = 6, letter 4 = 3)

Write the numbered letters to solve the puzzle.

Science Puzzle

The egg breaks and the chick

___ ___ ___ ___ ___ ___ ___.
1 2 3 4 1 5 6

Animal Life Cycles

Waiting to Hatch

Skill:

Read graphic information to answer questions

Chicks stay inside eggs for different amounts of time. The amount of time each chick stays in the egg depends on the size of the mother bird.

1. Which of the three chicks takes the longest to hatch? ____________________

2. Which takes longer to hatch, a chicken or a hummingbird? ____________________

3. Explain why a goose takes 14 more days to hatch than a hummingbird.

____

____

TARGETING SCIENCE YEAR 3 © PASCAL PRESS ISBN: 9781925726527

Skill:

Complete charts or graphs to demonstrate understanding of science concepts

Write to describe each stage of the life cycle.

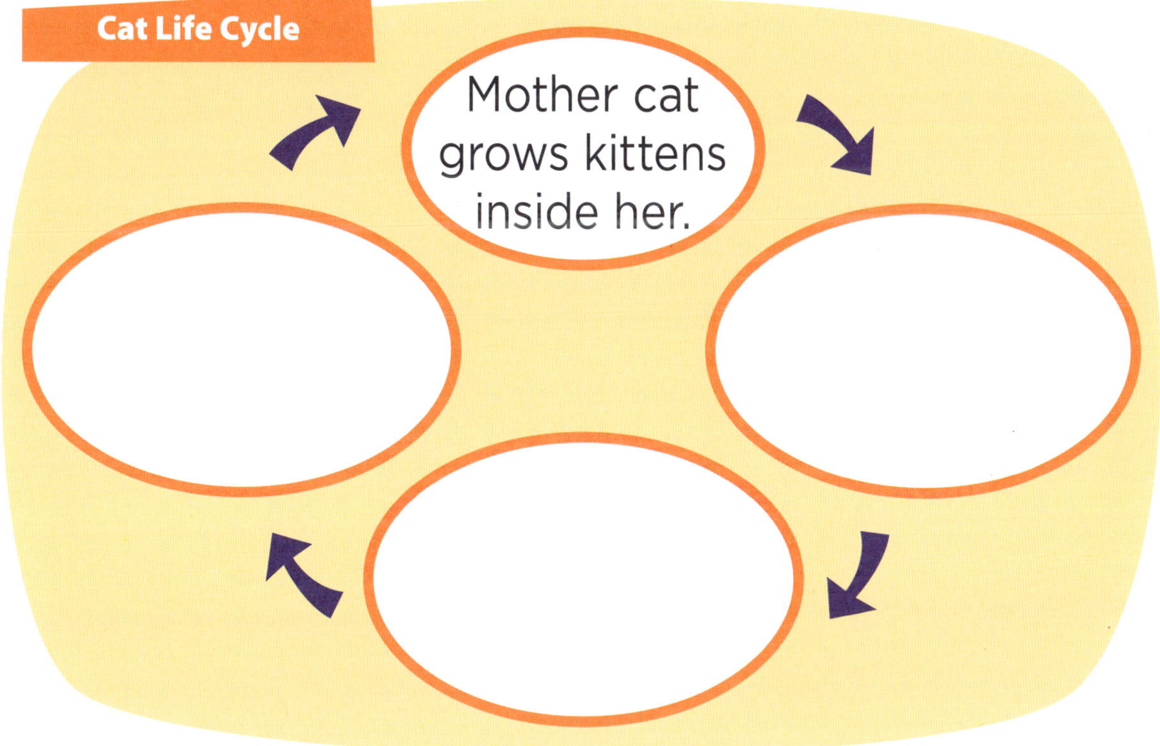

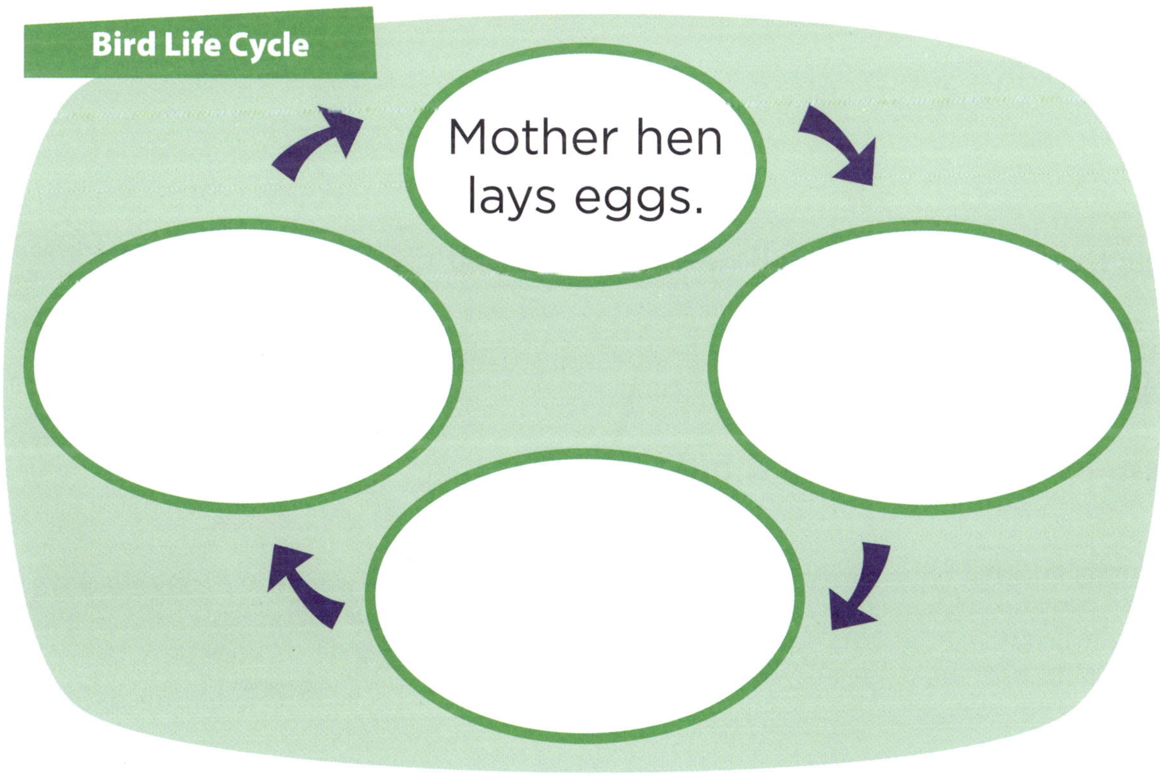

Review the Words

Skill:

Apply content vocabulary in context sentences.

Read each clue and write the missing word on the lines.

germinate pollen reproduce stamen

1. The pistil and the ___ ___ ___ ___ ___ ___ are parts of a flower.

2. When conditions are right, a tiny seed will

___ ___ ___ ___ ___ ___ ___ ___ ___.

3. Tiny grains of ___ ___ ___ ___ ___ ___ stick to an insect.

4. Flowers ___ ___ ___ ___ ___ ___ ___ ___ ___

by making seeds.

Write the letters from the yellow boxes to answer the riddle.

Science Riddle

Where does a tiny plant hide?

inside every ___ ___ ___ ___

Life Cycles

Bean Sprout in a Bag

Skills:

Gather and record scientific data from observation.

Follow a sequence of directions to complete an investigation.

When you plant a seed in soil, you cannot see it germinate. But here is an easy way to watch a bean seed germinate. Prepare to be amazed!

What You Need

- bean seed
- paper towels
- plastic sandwich ziplock bags
- masking tape

What You Do

1. Fold a paper towel to fit in a plastic sandwich bag. Sprinkle the paper towel with enough water to make it damp before you place it in the bag. It should not be dripping wet, just moist.
2. Place a bean seed on the paper towel and zip the bag.
3. Tape the bag in a warm place, such as on a window that gets some sunlight. Hang it so you can observe the bean inside.
4. Put together and hang two more bags in case the first one doesn't work. The beans should germinate in a few days.
5. Keep a journal of what you observe each day, beginning with today. If the paper towel dries out, add a few drops of water.
6. After about two weeks, plant the sprouts in a pot of soil, with their leaves above the soil. Or, if the weather is warm and sunny, plant them outdoors.

From Seed to Plant

Concepts:

Reproduction is essential to the continued existence of every kind of organism.

Plants undergo a series of orderly changes in their life cycles.

Organisms have structures and functions that help them survive in an environment.

A seed doesn't look like the plant it will become. But inside are the tiny beginnings of a **root** and a stem. The seed has a food supply, too. A seed coat covers and protects everything inside the seed.

When **conditions** are right, the seed will **germinate**, or grow. Warmth from the sun, water from the rain, and food from the soil make the right conditions for growth.

As a plant grows taller, it also **develops** new parts. Its new leaves help make food for the plant. New roots hold it in the soil. Many plants develop flowers.

Define It!

conditions: the weather, soil, and light affecting growth

develop: to grow into an adult

germinate: to begin to grow; to sprout

root: a plant part that grows down into the soil

Write the answer to each question.

1. What protects a seed? ______________________
2. What happens to a seed when conditions are right? ______________________
3. What new plant parts develop? ______________________

Life Cycles

TARGETING SCIENCE YEAR 3 © PASCAL PRESS ISBN: 9781925726527

Flowers and Their Seeds

Define It!

life cycle: the course of changes that happen during the life of a living thing

reproduce: to produce offspring

Concepts:

Changes that flowering plants go through during their life form a pattern.

Organisms have structures and functions that help them survive in an environment.

Almost all plants on Earth have flowers. Flowers are beautiful to look at and they smell good, but they also do important work for the plant. Flowers make seeds, which are needed for the plant to **reproduce**, or make new plants. Sometimes the seeds drop to the ground. Other seeds are carried away and dropped by animals or by the wind. Many flowering plants produce their seeds and then die. When the conditions are right, the seeds will produce new plants. A whole new **life cycle** will begin.

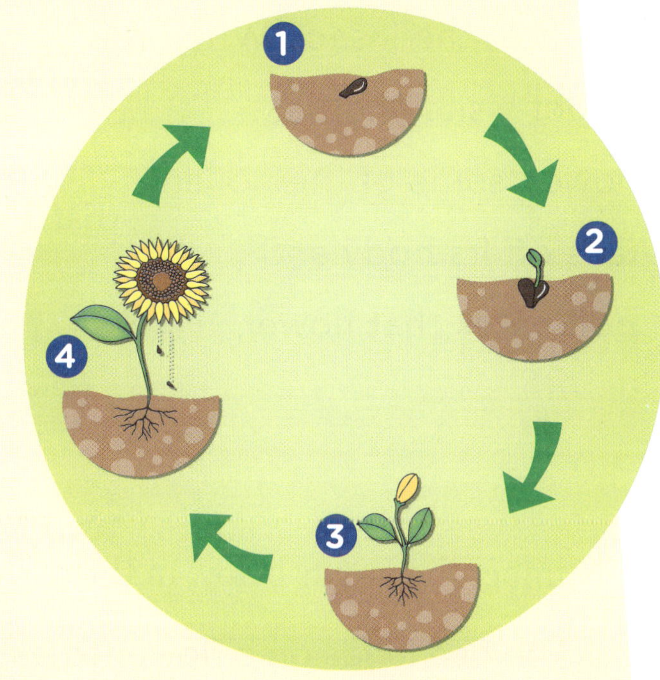

Write the missing words.

1. Seeds help a plant to make new plants, or ________________.

2. A plant germinates, grows and develops, reproduces, and dies. This is called its ________________.

3. Seeds can be carried by ________________ or ________________.

Life Cycles

Insects Carry Pollen

Concepts:

Plants undergo a series of orderly changes in their life cycles.

Organisms have structures and functions that help them survive in an environment.

Flowers make seeds when **pollen** travels from the **stamen** of a flower to the **pistil**. Insects such as bees, moths, and butterflies help carry pollen from one flower to another. When an insect lands on a flower, pollen sticks to the insect. When the insect visits another flower, some of the pollen falls off its body and **pollinates** that flower.

Define It!

pistil: the part of a flower in which seeds develop

pollen: tiny grains that help a plant produce seeds

pollinates: brings pollen to a plant

stamen: the part of a flower with pollen at the tip

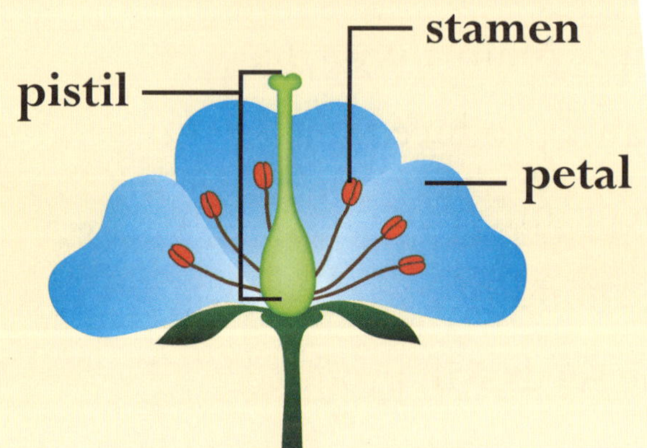

Number the events below in the correct order.

____ The bee lands on a flower to eat and drink.

____ The bee flies to another flower.

____ Pollen sticks to the bee.

____ The flower makes seeds.

____ Pollen falls off the bee.

Write the words to complete the sentence.

When a bee carries ______________ from one flower to another, the bee ______________ the flower.

Life Cycles

TARGETING SCIENCE YEAR 3 © PASCAL PRESS ISBN: 9781925726527

Some plants make seeds that have little hooks. These hooks can attach to animals passing by. Barley is a type of grass plant that moves its seeds in this way.

Look at each picture below. Then write the letter of the sentence that matches that picture.

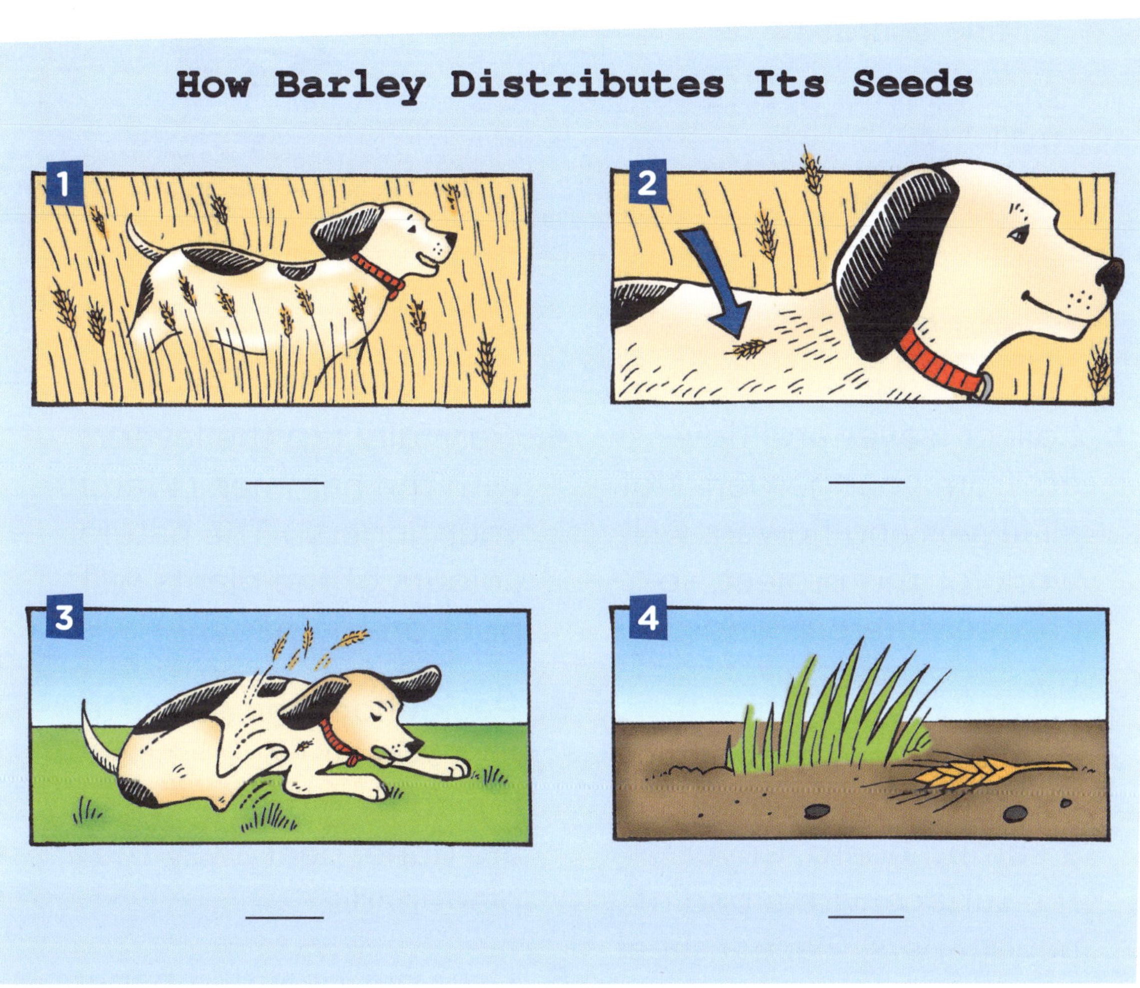

Skills:

Analyse and interpret information presented in a visual format.

Infer meaning from a sequence of pictures.

a. The dog scratches and sheds the seeds.

b. The dog rubs against the grass.

c. Seeds land on the ground and later grow into new plants.

d. Barley seeds attach themselves to the dog's fur.

Plant-Print Papers

Skill:

Follow a sequence of directions to complete a nature project.

What You Need

- leaves and flowers
- drawing paper
- paper towels
- cutting board
- small hammer or rubber mallet
- newspaper

What You Do

1. Collect leaves and flowers from your garden. Not all leaves and flowers will work for this project, so choose different kinds to try.
2. Find a surface that a hammer won't hurt, such as a cutting board. Cover the surface with a pad of newspapers. Place a sheet of drawing paper on top of the newspapers.
3. Arrange some plant parts on the drawing paper in a pretty design. Cover this with three layers of paper towels.
4. Carefully tap the layers with the hammer. (Watch your fingers!) The natural colours of the plants will print on the paper.
5. Peek under the paper towels to see if you want to tap some more. When you are done, peel away the plant parts.
6. Use your beautiful papers to make bookmarks or cards.

TARGETING SCIENCE YEAR 3 © PASCAL PRESS ISBN: 9781925726527

Life Cycle of a Plant

Skills:

Describe the life cycle of an organism.

Write explanatory text to convey information clearly.

Look at the diagram of the life cycle of a plant. Describe each step of the life cycle on the lines below.

1 ______________________________

____

2 ______________________________

____

3 ______________________________

____

4 ______________________________

____

Food for Growth

Concepts:

Living things need energy to grow and develop.

Animals get energy from food.

Living things are different from **non-living** things. One difference is that a living thing grows and changes, or develops. To do this, a living thing needs **energy**. Animals get their energy from food. Many young animals need help getting food, so their parents feed them. A kitten drinks milk from its mother's body. Robin **nestlings** open their beaks and their parents feed them. You need energy from healthful foods in order to grow and develop, just as other **organisms** do.

Define It!

energy: strength or power

nestling: a baby bird too young to leave the nest

non-living: not living

organism: a living thing

Write the answer to each question.

1. What do organisms need in order to grow and develop? ______________________

2. Where do living things get energy? ______________________

TARGETING SCIENCE YEAR 3 © PASCAL PRESS ISBN: 9781925726527

Fast and Slow Growth

Concept: Different types of animals grow and develop at different rates.

Define It!

hatchling: a baby animal that has just hatched

mammal: a warm-blooded animal

offspring: an animal's young

reptile: a cold-blooded animal

Rabbits are **mammals** that grow and develop quickly. A baby rabbit grows in its mother's body for about a month before it is born. A baby rabbit is called a *kit*. Rabbit kits are born without fur. The kits leave the nest when they are only two weeks old. By six months old, a rabbit is an adult and can start its own family. Rabbits live for about nine years.

Turtles are **reptiles** that grow and develop slowly for years. Some types of turtles grow for 10 years before having **offspring**, and others might grow for 30 years. Turtle eggs hatch in about two to three months. **Hatchlings** are on their own at birth. Their shells are soft at first and do not protect them. The hatchlings must scurry off to hide in a safe place. Some turtles live for 120 years or more.

Fill in the data.

1. Life cycle in years:

Rabbit ________________

Turtle ________________

2. Age when they leave the nest:

Rabbit ________________

Turtle ________________

Big Changes

Concept:

Some animals change entirely as they develop. This change is called metamorphosis

https://clickv.ie/w/9wgx

Use this QR code to access a video on this topic.

Define It!

chrysalis: a hard shell

larva: a wingless insect form

metamorphosis: a series of changes

moult: to shed an outer covering

nymph: a young insect stage

pupa: an insect stage after larva

Some animals make a very big change as they develop. This change is called **metamorphosis**.

Butterfly: Complete Metamorphosis

A butterfly develops in four stages: egg, **larva**, **pupa**, and adult. The egg is laid on a plant. A caterpillar, or larva, hatches from the egg. The caterpillar then becomes a pupa. The pupa sticks to a twig and forms a hard shell called a **chrysalis**. In the spring, an adult butterfly pushes out of the chrysalis.

A butterfly is coming out of its chrysalis.

Dragonfly: Incomplete Metamorphosis

A dragonfly develops in three stages: egg, **nymph**, and adult. Dragonflies lay their eggs in water. When a nymph hatches from an egg, it does not have wings yet. The nymph grows and sheds its skin. This is called **moulting**. The nymph moults many times before it grows wings and leaves the water to fly off.

A nymph is moulting.

Circle *true* or *false*.

1. Both butterflies and dragonflies moult. **true** **false**
2. Butterflies and dragonflies go through metamorphosis. **true** **false**

TARGETING SCIENCE YEAR 3 © PASCAL PRESS ISBN: 9781925726527

What's Inside an Egg?

Skill:

Analyse and interpret information presented in a visual format.

Chickens and other birds lay eggs in order to reproduce. A chicken egg holds the beginnings of growth.

Read about some of the parts of a chicken egg. Then label each part to complete the diagram.

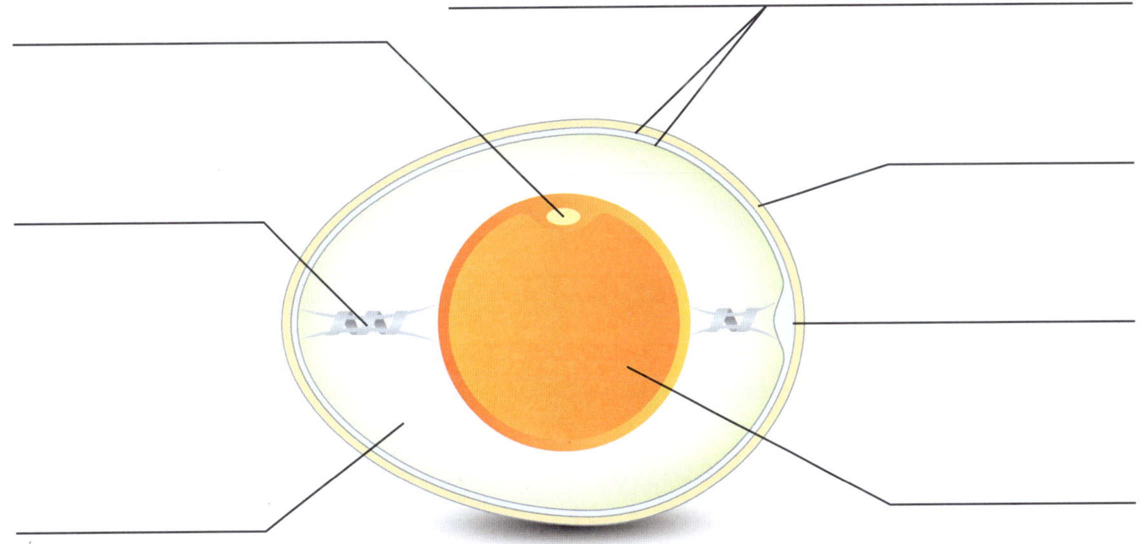

1. **Shell:** Hard outer covering that protects the egg. Water and air can pass through the shell.
2. **Inner and outer membranes:** Thin but strong layers that protect the egg.
3. **Air cell:** Empty space at the larger end of the egg, between membranes.
4. **Chalaza:** (kuh-LAY-zuh) Two ropes of egg white that hold the yolk in place.
5. **Albumen:** Clear liquid, also called the egg white, because it turns white when it is cooked.
6. **Yolk:** Yellow part of the egg. It is food for the developing chick.
7. **Germinal disk:** Small white circle on the yolk, which can develop into a chick.

Either/Or Questions

Skill:
Apply content vocabulary.

Write each answer.

1. Is an organism living **or** non-living? ____________________

2. Is a nestling a kitten **or** a baby bird? ____________________

3. Does energy come from food **or** a hard shell? ____________________

4. Is a rabbit a reptile **or** a mammal? ____________________

5. Are hatchlings a safe place **or** offspring? ____________________

6. Is metamorphosis a change **or** a chrysalis? ____________________

7. Is a chrysalis a hard shell **or** an adult insect? ____________________

8. Does a nymph stick to a twig **or** moult? ____________________

9. Is a reptile warm-blooded **or** cold-blooded? ____________________

10. Is a caterpillar a larva **or** a mammal? ____________________

TARGETING SCIENCE YEAR 3 © PASCAL PRESS ISBN: 9781925726527

I Have Grown

Skill:

Follow a sequence of directions to complete a chart of scientific data.

When you were born you probably measured about 50 cm long, as most babies do. How much have you grown since then?

1. Have someone measure your height with a measuring tape.
2. Fill in the data chart about yourself. (If you don't know how long you measured at birth, use 50 cm.)

Data About Me!

When I was born, I measured ______ cm long.

Today, I measure ______ cm tall.

I have grown ______ cm taller since I was a newborn.

My eye colour is ________________.

My hair colour is ________________.

I write with my ________________ hand.

I ________________ have freckles. (do/do not)

I ________________ wear glasses. (do/do not)

I Have Grown

Skill:

Follow a sequence of directions to complete a project.

What You Need

- a helper
- large sheet of butcher paper
- scissors
- crayons, pen-cils, markers, paint

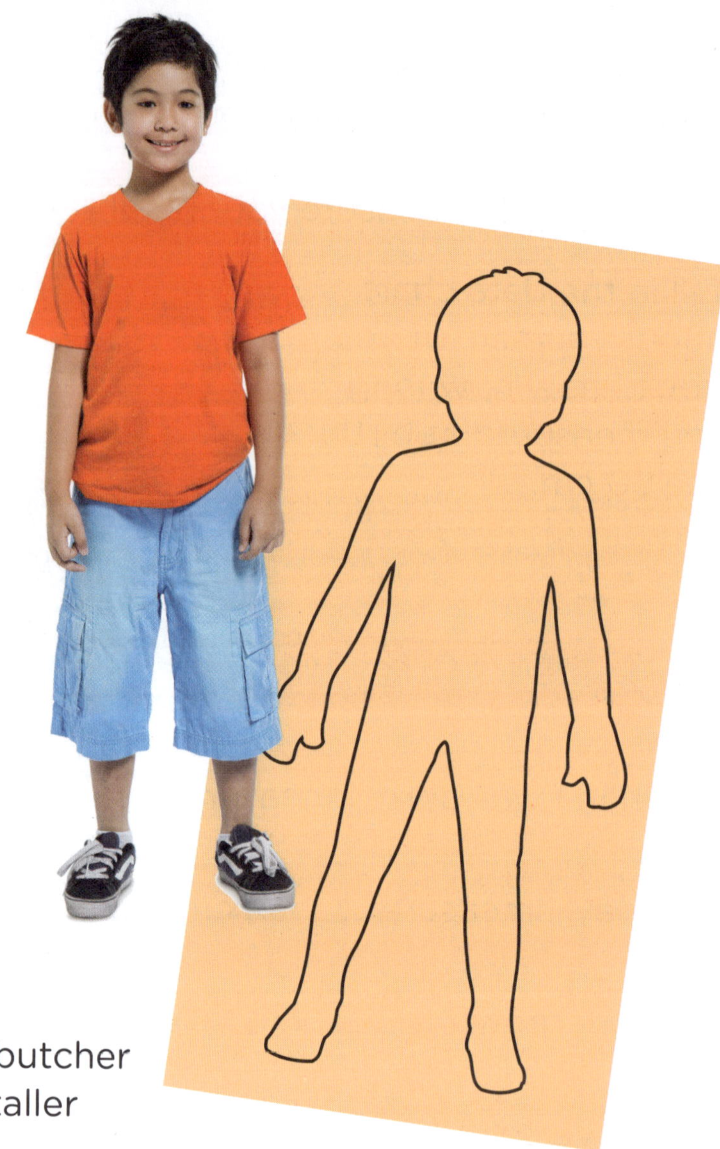

What You Do

1. Use a large sheet of butcher paper that is a little taller than you are.
2. Lie on the paper and have someone trace your body outline with a crayon.
3. Cut out the shape, leaving some paper at the bottom for the ground. Draw and colour your hair, clothing, shoes, and the ground with crayons.
4. Use a pencil to draw the features of your face. Then go over the pencil lines with paint or markers. Finish colouring in with crayons.
5. Hang your life-sized look-alike on your wall or door.

Skill:

Use information gained from words and illustrations to demonstrate understanding of text.

Read the clues about baby animals. Draw a line from the word to the picture. Then label the picture.

giraffe	human
duckling	sea turtle

Did You Know?

Hatchlings and newborns are able to do different things for themselves. Some need their parents to feed and care for them. Others are on their own right away.

1. These hatchlings scurry across the sand and into the sea. They must hide because their shells are soft and do not protect them.

____

2. A day and a half after they hatch, these little ones can find food and even swim. They can't fly yet, but their feathers grow in quickly.

____

3. This baby stands up within an hour of being born. It might be 1.8 m tall.

____

4. This baby cannot do much for himself. He won't walk for about a year, so his parents carry him.

____

From Egg to Frog

Concepts:

Many traits of an organism are inherited from its parents.

Some inherited traits appear as the animal grows.

Define It!

froglet: a tiny frog

gills: body parts of a water animal

hind: at the back

lungs: body parts for breathing air

tadpole: a newly hatched frog

Life Cycle of a Frog

1 A mother frog lays her eggs in water. Each tiny egg is in a floating ball of jelly.

2 A tiny **tadpole** pushes out. It has a long tail for swimming and **gills** for breathing.

3 The tadpole eats water plants. It grows **hind** legs and gets bigger.

4 The tadpole grows **lungs** for breathing, and loses its gills. It also grows front legs.

5 The tadpole has become a tiny **froglet**. It jumps onto land. The rest of its tail will soon be gone.

6 The froglet eats insects and grows into an adult.

Circle *true* or *false*.

1. An adult frog breathes with gills. **true** **false**
2. A tadpole eats water plants **true** **false**
3. A tadpole grows legs and starts to lose its tail. **true** **false**

TARGETING SCIENCE YEAR 3 © PASCAL PRESS ISBN: 9781925726527

Are You My Parent?

Draw a line to match the parent with its offspring. Use the list of traits to help you. Make a check mark by each trait as you make a match.

Did You Know?

Offspring resemble their parents. There are many kinds of apes and monkeys. Most monkeys have tails, while apes do not. The hair, face, colouring, and arms or legs of the animals help to tell them apart, too.

Skills:

Analyse and interpret information presented in a visual format.

Use a study checklist.

chimpanzee

orangutan

squirrel monkey

______ face

______ tail

______ colouring

______ ears

______ face

______ hair

______ colouring

______ arms

______ face

Life Cycles - a First Nations Perspective

One of the ways that First Nations Peoples distinguish seasons is based on their knowledge of animal life cycles.

Bininj and Mungguy Peoples of the Kakadu in the Northern Territory know that Magpie geese breed in the wet season when food is plentiful. This is a time when eggs can be collected. First Nations people know to only take the small amount that they need, so that the geese can continue to survive in that environment as they have done for thousands of years.

Hunting the magpie geese begins in wurrgeng (the cold time) when the magpie geese are fat and heavy after eating lots of food, and continues into gurrung (the hot, dry season).

The life cycles of many species of moths are very important to First Nations because of the nutritious food they provide at different stages of their life cycle.

There are four main stages in the life cycle of moths – egg, larva, pupa and adult. The adults mate and the female lays and egg. The egg then hatches into a larva, also called a caterpillar.

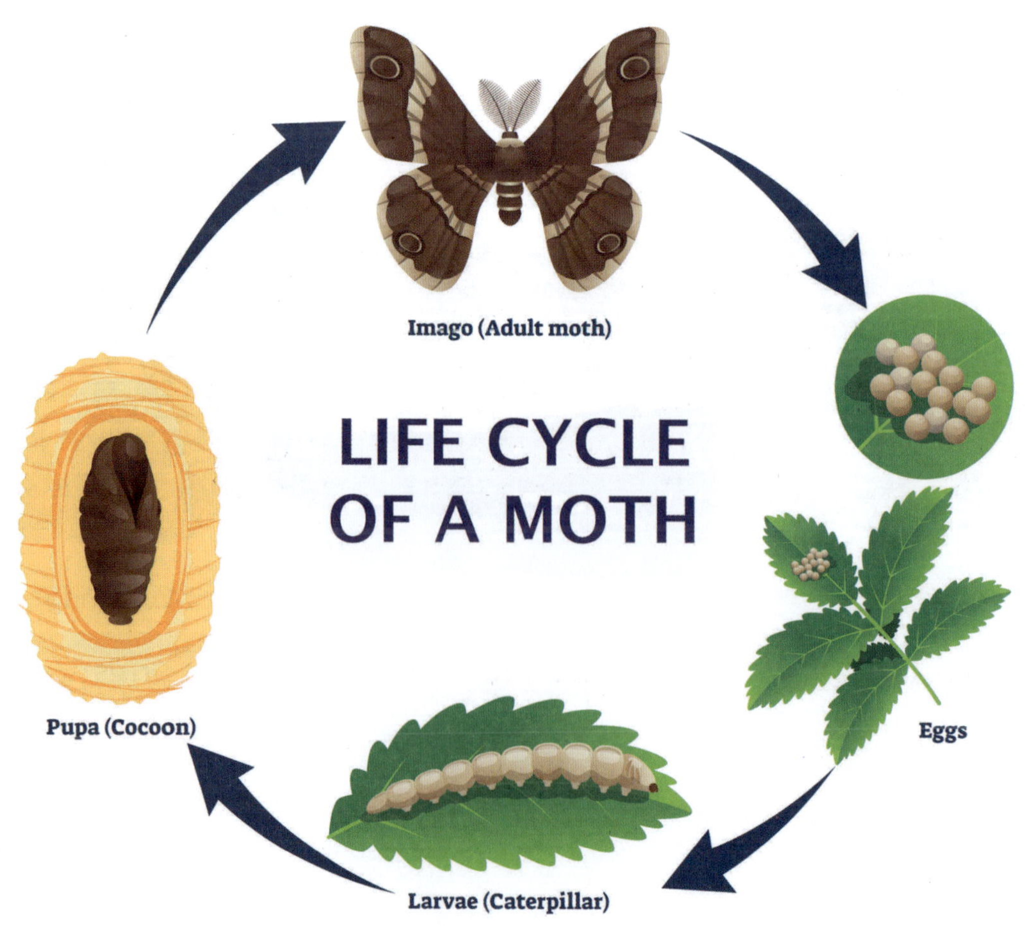

TARGETING SCIENCE YEAR 3 © PASCAL PRESS ISBN: 9781925726527

The caterpillar stage of a large wood moth, known to the Adnyamathanha Peoples of the Flinders Ranges region in South Australia as witjuri, burrows holes into the wood of Acacia to feed on plant sap. They look for sawdust piles at the base of the tree and harvest the larvae as a food source rich in protein.

The Gunditjmara Peoples of western Victoria dig at the bottom of gum trees to find the pupa stage of the moth in winter to roast in the ashes of a fire for food.

The bogong moth is found in huge numbers in the cool caves of the Snowy Mountains on the lands of the Ngarigo Peoples in southern New South Wales. The moths are very high in fat and are very nutritious. They are caught in nets using smoke, then gently cooked at the edge of a fire and the body part is eaten.

1. Would you ever be able to eat an insect?

____

2. Did you know that in many parts of the world insects are now being seen as a solution to the hunger in many poorer countries? Why would insects be a good solution for this?

____

Rock or Mineral?

Have you ever had a good look at the rocks around your school or in your garden? There is a lot more to them than you might think.

Rocks and minerals are non-living substances, meaning that they do not come from an animal or a plant.

A **mineral** is the raw material a rock is made from.

A **rock** is a solid made up of two or more minerals. People who study rocks and minerals (Geologists) classify rocks according to the way they are formed.

Have a look around outside and see if you can find about six different, small rocks and bring them inside to examine them more closely. If you can find a magnifying glass, it may help you.

Make some observations about your rocks in the table below:

Rock Number	Colours present:	Heavy or light compared to others same size?	Smooth or rough?	Are there crystals present?	Are there lines or layers?
1					
2					
3					
4					
5					
6					

Minerals are like the building blocks of rocks. The different characteristics of rocks depend on the minerals they are made from.
These rocks are different colours because of the different minerals that were present when they formed.

TARGETING SCIENCE YEAR 3 © PASCAL PRESS ISBN: 9781925726527

Types of Rocks

There are **three main types of rocks:**

Sedimentary Rocks

Sedimentary rocks are formed over many years out of sediments, or tiny pieces of rock and soil. The weight of the water and earth above them squeezes and cements them together. Streams or rivers will carry lots of sediment to a larger body of water and settle at the bottom. Millions of years later, they will form into solid rock.

Examples of sedimentary rocks:

- Limestone
- Coal
- Sandstone
- Shale

Igneous Rocks

Igneous rocks form when melted rock cools and hardens. Hot moulten rock called magma or lava from volcanoes cools down and hardens when it reaches Earth's surface or somewhere within the crust.

Examples of igneous rocks:

- Basalt
- Granite
- Quartz
- Pumice
- Tuff

Metamorphic Rocks

Metamorphic rocks are formed by huge amounts of heat and pressure. They are made from other types of rocks such as sedimentary or igneous rocks. They are formed deep under the surface of Earth, where the heat from Earth's core and the huge weight of the rocks and soil above cause them to melt and change.

Examples of metamorphic rocks:

- Gneiss
- Marble
- Slate
- Schist

Were you able to find any of these types of rocks in your collection?

The Rock Cycle

Rocks are always changing and the process of change is called the **rock cycle.** The changes happen very slowly (over millions of years). A rock can change from one type to another through this continuous movement on the surface of Earth.

1. What causes sedimentary rock to become a metamorphic rock?

____

2. Where might a metamorphic rock melt and come to Earth's surface?

____

3. How does metamorphic rock become sediment again?

____

Rocks and Minerals

 ISBN: 9781925726527

Uses of Rocks and Minerals

How many ways can you think of that people use rocks and minerals in their everyday life? Write as many ideas as you can here:

Did you realise rocks and minerals are used for these things?

Building materials: Many types of rock, mineral and stone are used to make building materials for houses, roads and high-rises. It's tough, durable, natural and often easy to source.

Cosmetics and jewellery: Minerals such as iron oxides, talc, zinc oxide, and titanium dioxide are ground into tiny particles to create makeup.

Cars: Cars are made from many different minerals. Their steel body is made from the iron-rich minerals like magnetite and hematite. Door handles and badges are often coated in chromium which comes from the mineral chromite.

Appliances and Health: Taps are made of stainless steel. The bathroom floor and the tiles above the bathtub are ceramic, also made from clay minerals. Copper wiring powers your appliances and the mineral fluorite in your toothpaste protects your teeth from decay!

Habitats for Wildlife: Materials found in the Earth's surface are the primary sources of minerals that are used to sustain plant life, which in turn provides food for animals. Rocks and soil provide habitats for millions of different species of animals and plants. Without them, there would be no life on Earth.

Footings for Giant Trees: Giant trees could not stay upright without the help of giant rocks and firm soil to secure their roots.

Rock and mineral detective:
Look around your home, classroom or school. How many uses of rocks and minerals can you find?

The Importance of Soil

Without **soil**, there would be no life on Earth.

Soil is made up from rocks, sand, clay, dead plants, animal remains and fungi.

Soil is incredibly important for these four main reasons:

Soil is vital for growing plants to feed animals.

Soil provides habitats for organisms.

Soil helps to clean our water by acting as a filter.

Soil is needed for plants to provide oxygen for animals

Soil is one of the world's most important resources, and yet it is often wasted or not cared for properly. It can take up to 1000 years for 3cm of soil to form on Earth's surface, as a result of the weathering and erosion of rocks.

Millions and millions of tonnes of soil are washed into the ocean every year, and a lot of that is due to the actions of people.

What can we do to stop soil being washed or blown away?

There are many things you can do in your school or garden to prevent precious soil from being wasted.

Prevent soil from being washed or blown away by building barriers.

Plant trees and grasses to hold soil in place, especially on slopes.

Reduce, reuse, recycle and repurpose, so that we have less rubbish going into our soil.

Stop chemicals and pollution from going into the soil.

Find somewhere in your garden or school where soil is not being looked after properly. Make a poster to show how it could be saved!

 TARGETING SCIENCE YEAR 3 © PASCAL PRESS ISBN: 9781925726527

Types of Soil

There are **four main types of soil** - Sand, loam, clay and silt. You can tell a lot about the type of soil by the way it feels.

Sandy Soil has large particles. It feels rough and gritty when you rub it, and it doesn't contain many nutrients. Water flows through sandy soil very easily because there are big spaces between the particles.

Silty Soil is smooth when you touch it. When you roll it between your fingers, dirt is left on your skin.
It holds water for longer.
It feels cold and drains poorly, so it is not great for growing plants. It cracks and crumbles when dry.

Clay Soil has the smallest particles found in soil. It does not let water through it easily.
It is sticky when it is wet, and smooth when it is dry.
In the summer, it could turn hard and compact.

Loamy Soil contains a good mix of the three soil materials - silt, sand, and clay.
It is best soil for growing plants.
It is dark in colour and soft, dry, and crumbly. It absorbs water, but drains well.
This is where you would find most types of underground worms and insects living.

A **geologist** is a scientist who studies rocks and soil. Geologists often use a '**key**', like the one on the next page and other equipment to help them to decide what type of soil they are working with.

Testing Soil Texture by Feel

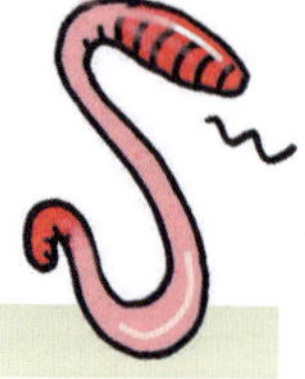

Place two teaspoons of soil in your palm, add water droplets and mix until it feels like play dough.

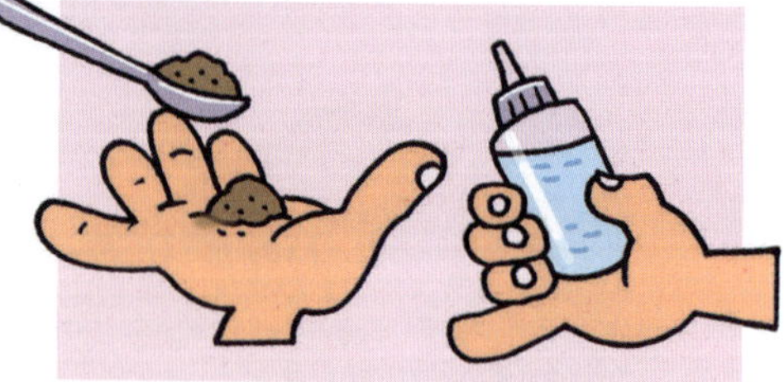

Does the soil stay in a ball when squeezed?

NO → This is SAND

Okay. Hold the ball between your thumb and forefinger and gently squeeze it upwards into a ribbon of even thickness. Keep going until it breaks from its own weight.

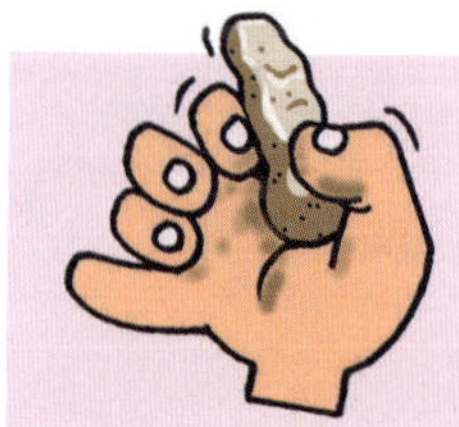

Does the soil stay like a ribbon?

NO → This is LOAMY SAND.

Does the soil make a short ribbon (20mm) before it breaks? — Or does the soil make a medium ribbon (40mm) before it breaks? — Or does the soil make a long ribbon (over 50mm) before it breaks?

Short ribbon (20mm):

Does the soil feel very gritty?

- YES → SANDY LOAM
- NO → LOAM or SILT LOAM

Medium ribbon (40mm):

Does the soil feel very gritty?

- YES → SANDY CLAY LOAM
- NO → CLAY LOAM or SILTY CLAY LOAM

Long ribbon (over 50mm):

YES

Does the soil feel very gritty?

- YES → SANDY CLAY
- NO → CLAY

Soil is made up of many layers called **horizons**. Each layer has its own characteristics that play an important role in what the soil is used for and why it is important. If you put the horizons together, they form a soil profile.

There are lots of places on Earth where not every layer is present. If the top layers (humus and topsoil) are taken away, plants cannot grow.

Humus is the top layer of soil that is made up of living and decomposed materials such as leaves, plants, and bugs. This is also known as the organic layer. A small patch of this soil (just 1 square metre) can hold billions of living things such as insects, spiders, worms, fungi, and bacteria. Plants grow well in this layer of soil.

The **topsoil** is made up of minerals and decomposed organic matter. It is fairly thin (12-24 cm thick). This soil is used for putting under grass.

The **weathered rock fragments** layer is made of clay, iron, and organic matter.

The **subsoil** layer is made up mostly of large rocks. Plant roots are not found in this layer. It is called parent material because the upper layers developed from this layer.

The **bedrock** is a mass of rock such as granite, basalt, quartzite, limestone, or sandstone. This is not soil. The bottom layer is several feet below the surface.

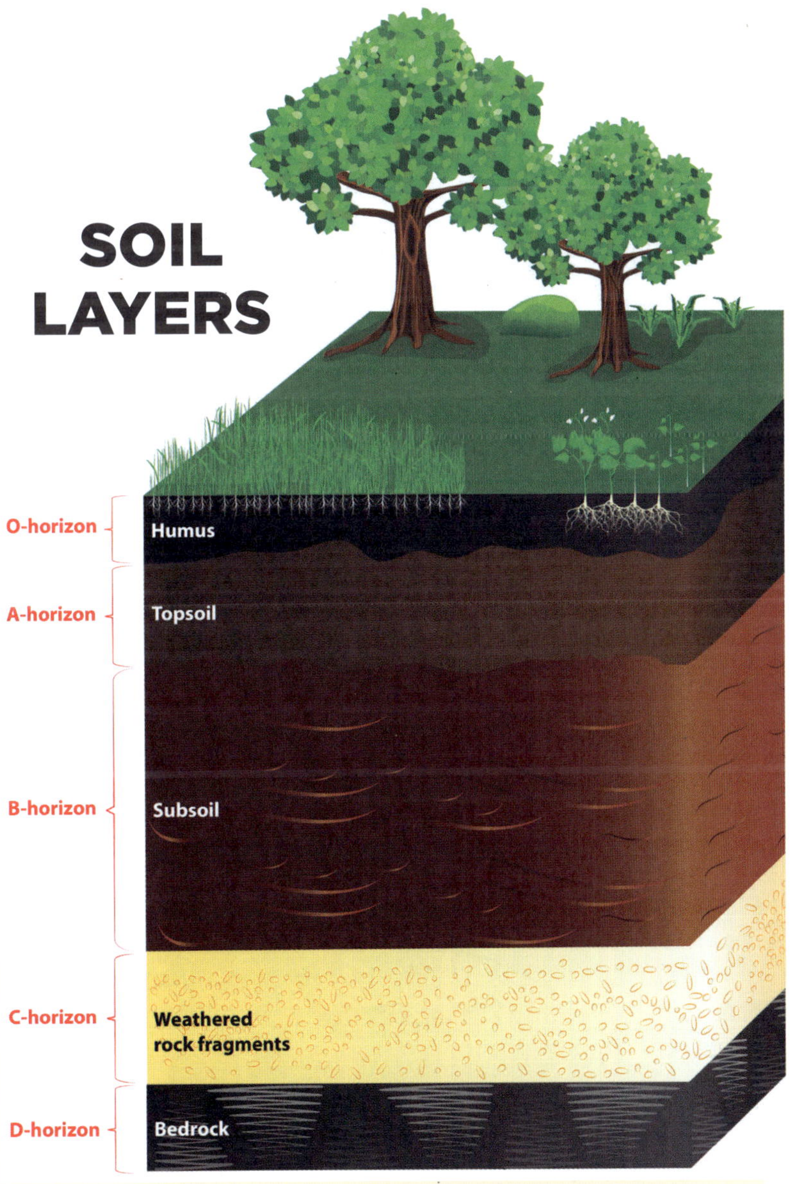

Rocks and Minerals - a First Nations Perspective

For thousands of years, First Nations peoples have used rocks and stones for making weapons and grinding tools. Their hard, durable properties made the perfect addition to axes, spearheads and other weapons and utensils. Their edges could be sharpened and ground to make thin blades or sharp points.

Rocks such as ochre are ground to a fine powder and mixed with water to make paint.

Grinding stones were used to grind and crush different materials. Bulbs, berries, seeds, insects and flowers were placed between a large lower stone and a smaller upper stone.

Flaked stone tools were made by hitting a piece of stone, called a core, with a 'hammerstone', which was often a pebble. This would cause the rock to 'flake' leaving a sharp edged rock that could then be fastened to timber using resin or other plant and animal materials.

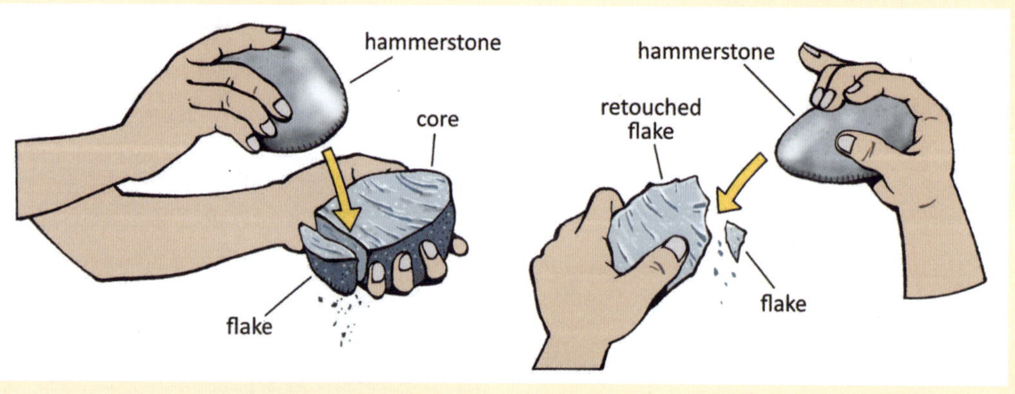

Look at these sharpened rocks. Can you make them into weapons by drawing and labelling the other parts you would like to add?

Fossils

Define It!

decay: to break down or rot

fossil: remains or marks of a living thing that lived long ago

mineral: something found in nature that is not plant or animal

sediment: soil or sand that forms layers on land or under water

A **fossil** is the remains or marks of a plant or an animal that lived long ago. Living things **decay** when they die. But if their bones, shells, or teeth are buried quickly by **sediment**, a fossil can form. Over millions of years, heavy layers of sediment pile up and harden into rock.

A *mould* is one kind of fossil. It is made when a plant or an animal rots away and leaves only its shape in the rock.

A *cast* is another kind of fossil. It is made when **minerals** fill a mould in the same shape as the plant or animal.

A *trace fossil* is made from things such as footprints or nests. They tell scientists how the animals moved and lived.

Check the box that answers the question.

1. Which of these would **not** make a trace fossil?
   - ☐ footprint
   - ☐ bone
   - ☐ nest

2. Which of these could show the entire shape of an animal?
   - ☐ sediment
   - ☐ trace fossil
   - ☐ cast

The Fossil Record

Concepts:

Earth's layers of rock comprise the fossil record.

The fossil record includes plants and animals that no longer exist anywhere.

Define It!

extinct: no longer to be found living

fossil record: the history of life on Earth as told by fossils

marker fossil: fossil that marks a time in the fossil record

trilobite: an extinct sea animal

The layers of rock on Earth have built up over time. Scientists try to understand what Earth was like long ago by studying fossils in the different layers of rock. Scientists call Earth's layers of rock the **fossil record**.

Trilobite fossils are common in the fossil record. Trilobites were sea animals that lived about 250 million to 540 million years ago. They are now **extinct**, or no longer to be found. These extinct animals are useful to scientists today. Trilobites are **marker fossils**. When a new kind of fossil is discovered, scientists want to figure out when that plant or animal lived. The fossil record can tell scientists if the organism lived before, after, or at the same time as trilobites.

Write the missing words.

1. Trilobites are ____________________ sea animals that had three main body parts.
2. The layers of rock in the ____________________ give clues about when an organism lived.

 TARGETING SCIENCE YEAR 3 © PASCAL PRESS ISBN: 9781925726527

Earth's Past

Define It!

continents: the seven largest bodies of land on Earth

fern: a feathery plant

landmass: a large body of land

The story of Earth's past tells of big changes over millions of years. Fossils are clues that tell how life on Earth changed over time. Fossils of a **fern** plant gave scientists a clue. The same fossils were found on different **continents**. The scientists began to think that the continents were once connected. Now scientists think that all the continents were one giant **landmass** about 200 million years ago. They call that landmass Pangaea (pan- JEE-uh).

Today, millions of years after Pangaea broke apart, the continents are still moving very slowly. The movement of continents builds mountains. When two continents push into each other, their rock layers push together and make a mountain. Sometimes the layers are pushed up all the way from the ocean floor. If those rock layers hold fossils, the fossils travel up with the rocks. This is why fossils of sea animals have been found at the tops of mountains.

Concepts:

We learn about Earth's early environments from fossils of plants and animals that lived long ago.

Some changes in Earth's surface happen slowly over millions of years.

Circle *true* or *false*.

1. Earth has always looked the way it does now. **true** **false**
2. Fossils can tell us what Earth looked like long ago. **true** **false**
3. Sea animal fossils cannot be found on a mountain. **true** **false**

Stuck in Time

Skills:

Infer meaning from pictures.

Classify events in sequential order.

Scientists have been digging up thousands of fossils in the city of Los Angeles, California! The area is called the La Brea (lah BRAY-uh) Tar Pits. Scientists have found the bones of mammoths and sabre-toothed cats there. These extinct animals were trapped in the tar pits about 30,000 years ago.

Write the letter of the sentence that matches each picture.

a. A hungry sabre-toothed cat attacked the mammoth.

b. Over time, both animals were buried in the tar pit.

c. A mammoth wandered into a pool of sticky tar.

d. The sabre-toothed cat was stuck, and died of hunger.

 TARGETING SCIENCE YEAR 3 © PASCAL PRESS ISBN: 9781925726527

Fossils Crossword Puzzle

Skill: Apply content vocabulary.

Use the vocabulary words to complete the crossword puzzle.

decay	fossil	trilobite	landmass
continents	extinct	sediment	fern

Across

4. the seven largest bodies of land on Earth

6. soil or sand that forms layers on land or underwater

8. no longer found to be living

Down

1. remains or marks of an organism that lived long ago

2. a large body of land

3. a feathery plant

5. an extinct sea animal that had three main body parts

7. to break down or rot

My State's Fossil

Skills:

Conduct research on a science topic and write an illustrated report.

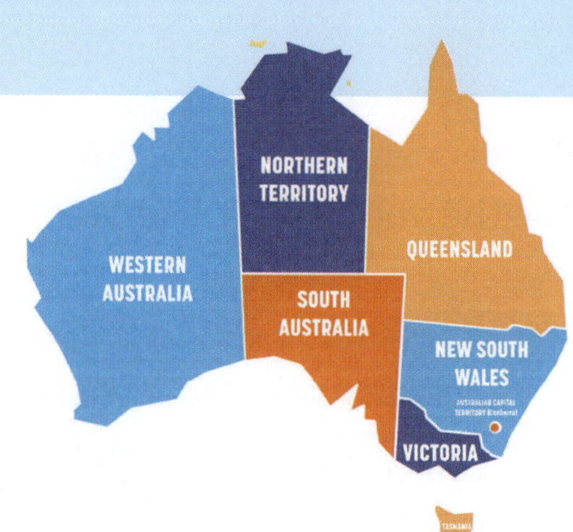

Several Australian states and territories have chosen a state fossil. Look for your state's fossil in a library book or on the Internet. If your state does not have a fossil yet, choose one that you think it should have.

Draw and write about your state fossil.

My state/territory: ______________________________

My state/territory's fossil: ______________________________

Draw

What makes this fossil a good fossil for your state?

____

____

____

____

____

Fossils

TARGETING SCIENCE YEAR 3 © PASCAL PRESS ISBN: 9781925726527

Make Your Own Fossil

Skill:

Follow a sequence of directions to complete a science project.

What You Need

- petroleum jelly
- plaster of Paris
- water
- bowl and spoon
- cardboard
- leaves and shells

What You Do

1. Coat a leaf and a shell with petroleum jelly. Set them aside.
2. Follow the directions to mix the plaster and the water. Have an adult help you.
3. Spread some plaster on the cardboard. Use enough plaster so that you can press the leaf and the shell into it.
4. Press the coated leaf and shell gently into the plaster. Try not to move them once they are in the plaster.
5. Let the plaster dry overnight. When it is dry, carefully remove the leaf and the shell.

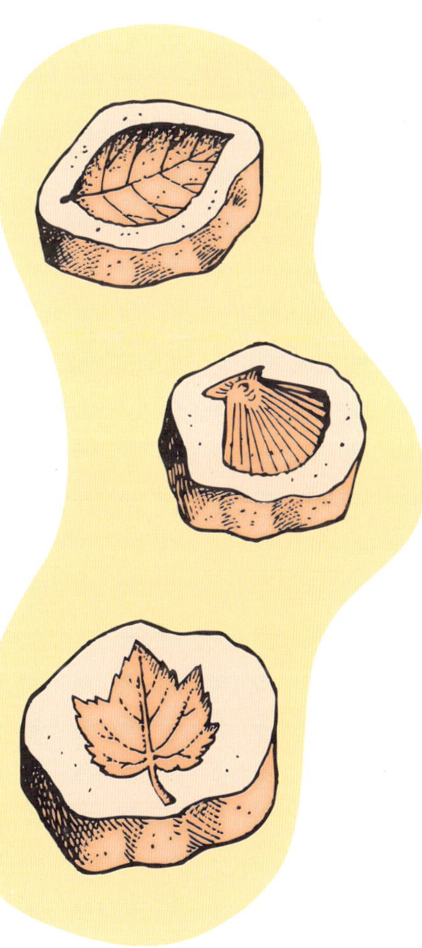

Fossils

Fossil Discovery

Skill:

Apply scientific knowledge to write a narrative about an imagined experience.

Pretend that you are a shell fossil. You are the remains of an extinct sea animal. Tell the story of how you came to be at the top of a mountain.

Hint

Layers of sediment built up on the ocean floor for millions of years. When the layers were pushed up by the moving continents, they became mountains.

 TARGETING SCIENCE YEAR 3 © PASCAL PRESS ISBN: 9781925726527

Concept:
Different forces shape the land.

Earth's **surface** changes over time. It takes many years and different forces to cause these changes.

Erosion is a force that shapes the land. Erosion happens when rock and soil are carried away by water and wind. For example, a stream can carry away rock and soil and carve a path in the land. Over time, erosion might turn that path into a canyon. That's how the Grand Canyon in the United States was formed!

Weathering is also a force that shapes the land. Weathering happens when water or wind break down or wear away rock. Have you ever found a smooth rock on the beach? That is an example of weathering.

Weathering and erosion are two natural forces that change the land. But people can change the land, too. Walking or riding along the same path wears the land down. Building roads changes the land, and so does planting crops and trees.

In nature, things are always changing. Water, wind, and people cause some of these **changes** to happen.

Skill:
Apply science vocabulary in context

Read the clue. Write the word.

1. This part of Earth changes over time.

___ ___ ___ ___ ___ ___ ___
1 5

2. This carries rock and soil away.

___ ___ ___ ___ ___ ___ ___
2

3. The wearing away of rock is called

___ ___ ___ ___ ___ ___ ___ ___ ___ ___.
3

4. Nature doesn't stay the same; it

___ ___ ___ ___ ___ ___ ___.
4 6

Write the numbered letters to solve the puzzle.

Science Puzzle

The land is changed by different

___ ___ ___ ___ ___ ___.
1 2 3 4 5 6

Geology

What Happened?

Skill:

Interpret information from graphic images

This is a picture of a road that is being built. People had to change the land to make room for the road. What are some things you think people did to change the land? Choose all that apply.

- ☐ Cut down trees
- ☐ Planted more trees
- ☐ Dug up big rocks
- ☐ Dug a big hole
- ☐ Made the land flat

Geology

TARGETING SCIENCE YEAR 3 © PASCAL PRESS ISBN: 9781925726527

Skill:

Conduct experiments and answer questions

Rocks in a stream become weathered when they are tumbled by the water. Here is how you can weather rocks yourself.

What You Need

- handful of gravel
- 2 small jars with tight lids
- 2 coffee filters
- strainer
- bowl
- water

What You Do

1. Place half the gravel in each jar, and fill the jars with water.
2. Shake one jar as many times as you can in 2 minutes. Do not shake the other jar.
3. Place a coffee filter inside the strainer. Hold it over the bowl. Pour everything from the jar you shook into the filter. Let the water drain.
4. Take the rocks out of the filter, but leave the sand. Then lay the sandy filter flat to dry.
5. Repeat Steps 3 and 4 with the unshaken jar.

Why did one filter have more sand?

____

____

____

Soil Erosion

Skill:

Conduct experiments and record results

Explore how wind and water move soil.

What You Need

- disposable plate
- dry soil or sand
- piece of wood
- cup of water

What You Do

1. Work outside (in the school playground, garden or at the beach) for easy cleanup.
2. Sprinkle some soil (or sand) on the plate.
3. Blow on the soil. What happened to it?

   ____

4. Now use the wood to prop up the plate.
5. Pour water on the soil. What happened to the soil?

   ____

   ____

6. How might you keep the soil from being blown or washed away?

   ____

   ____

Geology

TARGETING SCIENCE YEAR 3 © PASCAL PRESS ISBN: 9781925726527

Erosion, Weathering, or People?

Skill:

Label graphic images to demonstrate understanding of science concepts

What made the change? Write *erosion*, *weathering*, or *people*.

Geology

TARGETING SCIENCE YEAR 3 © PASCAL PRESS ISBN: 9781925726527

Dense Rocks

Concept:

Rocks from deep within Earth give scientists evidence of the composition of the planet's interior.

Our Planet's Interior

Define It!

dense: closely compacted; thick

elements: the basic chemical substances from which all matter is made

interior: situated within or inside

For thousands of years, people have wondered what is inside Earth. At one time, there were people who claimed Earth was hollow. Some thought the **interior** of Earth was full of holes like Swiss cheese. Still others believed it was the "underworld," where spirits of the dead went! But scientists eventually discovered that Earth's interior is made up of **dense** volcanic rock. In fact, a rock from inside Earth could weigh more than three times as much as one of the same size on the surface.

As Earth was formed, heavier **elements** such as iron and nickel sank to the centre of the planet. The less dense elements, such as oxygen and silicon, remained near the surface. However, over millions of years, some iron-rich volcanic rocks have made their way to the surface from deep within Earth. These rocks provide scientists with direct evidence of the types of materials Earth's interior is made of.

Meteorites are made of iron and nickel, just like the volcanic rocks from the core of our own planet.

Write *true* or *false*.

1. Earth's surface is made up of iron and nickel. __________
2. As Earth formed, heavier elements sank to its centre. __________
3. Earth's interior is made of dense rock. __________
4. Oxygen and silicon are more dense than iron and nickel. __________

TARGETING SCIENCE YEAR 3 © PASCAL PRESS ISBN: 9781925726527

Scientists figured out what the planet's interior was made of by studying volcanic rocks. By studying earthquakes, scientists were able to learn more about the **structure** of Earth. Earthquakes produce **seismic waves** that travel through the planet. The waves travel faster through dense material. And some waves cannot travel through liquid.

Define It!

seismic waves: vibrations that travel through Earth as a result of an earthquake

structure: the arrangement of the parts of a substance

How did scientists figure this out? Following an earthquake, geologists would measure how long it took seismic waves to be felt at different locations around the world. Based on the time it took for the waves to arrive, they concluded that Earth consisted of different layers—some more dense than others. And because some waves never arrived at all at some locations, the scientists realised that at least part of Earth's interior was made of liquid.

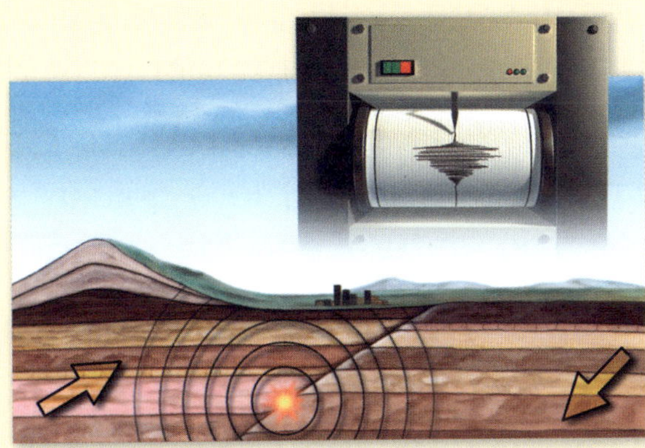

Concept:

By measuring seismic waves, geologists were able to learn more about the structure of Earth's interior.

Answer the questions.

1. What did scientists study to find out what Earth was made of? ____________________

2. What did scientists study to find out about the structure of Earth? ____________________

3. What two things did scientists conclude about Earth's interior after measuring seismic waves?

____

____

____

Places Where We Find Heat

Heat is all around us. We feel it when we walk into a warm room, touch a warm cup, stand in the sun or take a long shower. So, what is heat, and where does it come from? Heat is a form of energy.

Everything on Earth is made of particles, or bits, called **molecules**. Molecules are always moving. As an object is heated, the molecules in the object start to move faster. When the molecules are moving faster, the object starts to get warmer as its temperature rises.

When a pot on a stove starts to heat up from the electricity under it, the molecules move more quickly and the food feels warmer. Heat can come from natural and man-made sources.

Natural heat sources are the ones that are found in nature. These things don't need people to switch them on to make them hot.

Man-made heat sources are not found in nature. They need people to do something to them to make them hot.

Write NATURAL or MAN-MADE under each source of heat.

Heat on the Move

Define It!

molecule: the smallest particle of matter

particle: a very small piece or speck

temperature: a measure of how hot or cold something is

https://clickv.ie/w/Mxgx

Use this QR code to access a video on this topic.

Concepts:

Heat energy flows from warm objects to cool ones.

When matter heats up, its molecules move faster.

Have you ever warmed your hands by the fire? Then you know something about heat energy. Heat from the fire flowed to your cold hands and warmed them up.

Heat energy moves between things that have different **temperatures**. What happens when an ice cube is added to a bowl of hot soup? Coldness from the ice cube does not flow to the soup. The soup cools because heat from the soup flows to the ice cube. Heat energy flows from warm objects to cool ones.

Matter is made up of tiny **particles** called **molecules**. When matter heats up, its molecules move faster. Temperature is a measure of the energy of this motion. Temperature tells how hot or cold something is.

Complete the sentences.

1. Heat energy flows from things that are ________________ to things that are ________________.
2. When an object heats up, its molecules ________________.

Molecules in Motion

Concepts:

Conduction is the transfer of heat when there is a difference in temperatures between objects.

A conductor is matter that transfers heat easily, such as metal.

Define It!

conduction: the transfer of heat when there is a difference in temperatures

conductor: matter that provides an easy path for the flow of heat or other energy

transfer: to move from one place to another

Conduction is one way that heat energy is **transferred**. Conduction happens when two things are touching each other. For example, a metal spoon grows warm in a cup of hot tea. Why does this happen? When molecules bump into each other, heat energy is transferred. The heat moves from the faster-moving molecules of the hot tea to the slower-moving molecules of the cold spoon. The slower-moving molecules gain heat energy and speed up. Heat energy is transferred from one molecule to another until all the molecules in the tea and the spoon are moving at the same speed. The temperature of the tea and the temperature of the metal spoon become the same.

Metal is a good **conductor** because heat energy moves quickly through it. Cooking pots are metal because they quickly transfer heat energy from a stove to the food.

Answer the questions.

1. When the temperature of a cup of tea and the temperature of a spoon are the same, what do you know about their molecules?

____

2. What kind of heat transfer happens between food and a metal pan? ____________________

TARGETING SCIENCE YEAR 3 © PASCAL PRESS ISBN: 9781925726527

Define It!

convection: the transfer of heat energy, causing movement in liquids and gases

current: a flow

radiation: the transfer of heat energy in waves

solar: from the sun

Concepts:

Convection is the transfer of heat through a liquid or gas, causing currents.

Radiation is the transfer of heat energy through waves.

Convection is a second way that heat energy is transferred. Convection happens in liquids and gases, causing *convection currents*. Think of a pan of water on the stove. When the liquid gains heat energy, its molecules move faster and begin to spread out. When this happens, the heated liquid begins to rise. This causes **currents**, or flow, in the liquid. Currents formed in this way are called convection currents.

Radiation is a third way that heat is transferred. The sun is an example of radiation. The sun and other stars are always giving off energy. **Solar** energy travels to Earth through empty space in waves. Light waves, radio waves, and microwaves are some types of radiation from the sun.

Circle the answers.

1. Molecules of a liquid spread out when they are heated. **true** **false**
2. Convection currents may form in liquids and gases. **true** **false**
3. Radiation does not transfer heat. **true** **false**

Heat and Weather

Skill:

Interpret information gained from text and apply it to a diagram.

The sun's **radiation** warms the land on Earth. Then heat transfers from the land to the air by **conduction**. The air becomes warmer and lighter, so it rises. **Convection currents** begin to form. As the air goes higher into the sky, it begins to cool. When the air grows cool, it becomes heavier and sinks back toward land. Convection currents can cause breezes, winds, thunderstorms, and even hurricanes.

Label the diagram using the words *radiation, conduction,* and *convection currents.*

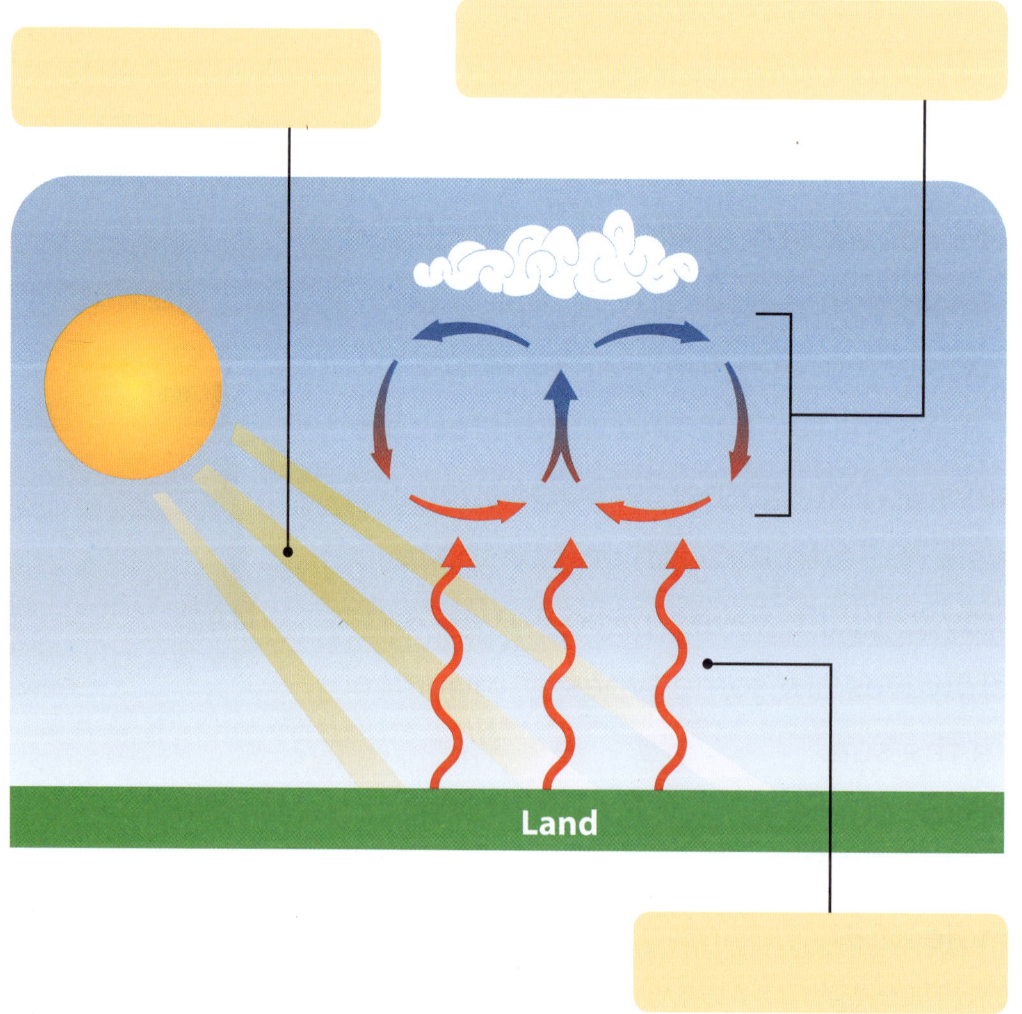

TARGETING SCIENCE YEAR 3 © PASCAL PRESS ISBN: 9781925726527

Either/Or Questions

Skill: Apply content vocabulary.

Write each answer.

1. Are molecules particles of matter **or** empty space?

   ____

2. Is temperature a measure of how far away something is **or** how cold it is? ______________________

3. Does the sun's energy reach Earth by conduction **or** radiation? ______________________

4. Do currents cause breezes **or** radio waves?

   ____

5. Is metal a conductor **or** a current? ______________________

6. Does solar energy come from liquids **or** from the sun?

   ____

7. Does convection make currents **or** particles?

   ____

8. Does conduction happen when temperatures are the same **or** when they are different? ______________________

Purple Swirl

Skills:

Plan and implement descriptive investigations to solve a specific problem in the natural world.

Answer questions and make inferences from observations.

To see the effects of heat transfer in action, all you need is food colouring and water. You can make the water swirl without even using your hands. Watch as heat energy does all the work!

What You Need

- clear plastic tray or rectangular container
- large disposable cup filled with hot water
- large disposable cup filled with ice water
- red and blue food colouring

Directions

1. Fill the tray with water that is almost warm.
2. Set the tray on top of the two cups, with the hot cup under one end and the cold cup under the other.
3. What do you think will happen when you add the dye? Why?
4. Add several drops of red food colouring to the water above the hot cup.
5. Add several drops of blue food colouring to the water above the cold cup.
6. Observe the movement of the colouring.

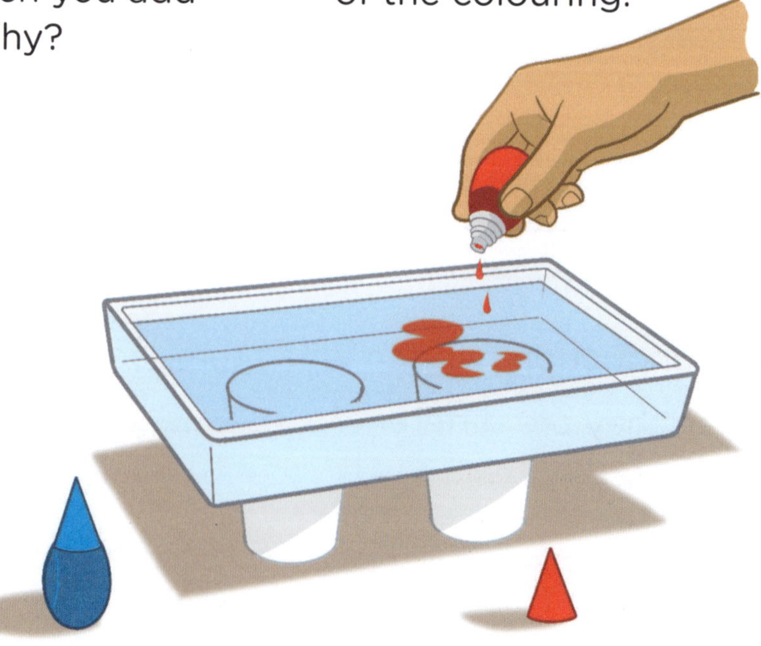

Heat Energy

TARGETING SCIENCE YEAR 3 © PASCAL PRESS ISBN: 9781925726527

What Did You Discover?

1. In what direction did the warmer red-coloured water move?

____

____

2. Use what you know about the transfer of heat energy to explain why the water did this.

____

____

____

3. What were the swirls of colour that you observed?

____

____

4. What happened to the colours when you waited awhile longer?

____

____

5. What do you think is true about the temperature of all the molecules in the purple water?

____

____

Apply What You Learned

Skills:

Apply scientific knowledge to writing a narrative about an imagined experience.

Use descriptive details and clear event sequences.

Pretend that you are a molecule of water. One day, you are put into a soup pot on the stove. What happens next? Tell your story.

Hint

Heat energy makes molecules move faster and spread out. Convection currents begin to form when heated liquid rises.

TARGETING SCIENCE YEAR 3 © PASCAL PRESS ISBN: 9781925726527

Using Thermometers

https://clickv.ie/w/Vygx

Use this QR code to access a video on this topic.

There are many different types of **thermometers**, but they all have one purpose: to measure changes in **heat** energy.

Read each of these uses for a thermometer and match them with a line to the thermometer that has been designed especially to do the job.

This thermometer is used to check if your meat is cooked in the oven.

This thermometer is used to check that the water in your fish tank is not too cold.

This thermometer is used to check a person's temperature without touching them.

This thermometer is designed to measure the temperature of a room

Vets use this thermometer to measure the temperature inside pets.

This thermometer is used to measure the temperature inside a fridge.

This thermometer is used to measure the temperature of hot drinks.

A doctor might use this thermometer in your ear to check your temperature

Scientists use this thermometer for measuring temperatures in an experiment.

Insulators

When something is **insulated**, it is designed to keep **heat** in or out. Rooves of houses are insulated to keep the heat from the sun entering the house. Refrigerators are insulated to keep the cold air in.

Ice Cube Insulator Challenge:

Focus Question: Which material is the best insulator for an ice cube?

What You Need

- 3 ice cubes that are the same size and shape. Keep them in the freezer until you are ready to do the experiment.
- A piece of aluminium foil, a piece of paper and a tissue that are exactly the same size.
- A plate to place them on.

Directions

1. Place each material on the table in front of you.
2. Take the ice cubes from the freezer and wrap each ice cube in a different material. If you use tape, use the same small amount on each material so that it is a fair test.
3. Leave the wrapped ice cubes on the table for five minutes.
4. Predict what you think might happen. Why?
5. Unwrap each ice cube and check to see how much it has melted.

Record your results

1. How much did each ice cube melt?

Ice cube 1: Paper	Ice cube 2: Tissue	Ice cube 3: Foil

2. What happened? ______________________________

____

____

3. Which material was the best insulator? ______________________________

____

TARGETING SCIENCE YEAR 3 © PASCAL PRESS ISBN: 9781925726527

First Nations Perspectives - Keeping Warm

First Nations Australians had many ways of keeping warm, and one of these methods was making clothing. They made cloaks from the skins of animals such as quoll, platypus, sugar gliders, possums, kangaroos and emus. These skins were warm, and kept the people dry.

Some of these cloaks were long enough to cover a person from their neck all the way down to their feet. The type of cloak or clothing that was made depended on where the people lived, and the animal skins that were available.

To make the cloak, the animal was skinned, and then the skin was attached to a flat surface and held in place with either wooden pins or echidna quills. All of the flesh was scraped off and eaten, before the skin was allowed to dry. Nothing went to waste.

The cloaks were then decorated with art to make it look more beautiful. They used mussel and oyster shells to cut patterns in the leather and stone tools to beat it and make it softer and more comfortable.

The cloaks were then painted using ochre and black pigment. The artwork often included symbols of the person's identity and Country.

In some parts of Australia, it was traditional for First Nations people to be buried in their cloak with their special things.

Solid or Liquid?

Look around where you are right now. **Everything** you see is made of **matter**. Matter is anything that has weight and takes up space. Molecules are the tiny parts or bits that make up all matter. There are millions of molecules in a tiny pin head.

Your shoes, your desk, your hair and even the food in your tummy is made of matter (which is made up of these tiny molecules).

Matter can be broken into three main groups: Solids, Liquids and Gases.

Let's take a closer look at solids and liquids:

Solids

- Solids **hold their shape.** When you leave them alone, they stay in the same shape their were before. They only change shape if you do something to it, like cut it or melt it.
- Solids **do not flow.**

Liquids

- Liquids **take the shape of the container** they are in.
- Liquids **flow.** Some liquids flow slowly, like honey, and others flow quickly, like water.

Make two lists of all of the things that you can find around you that are either a solid or a liquid. You might need to go for a bit of a walk to fill your lists.

Solids	Liquids
How did you know these were solids?	How did you know these were liquids?

1. Was it easier to find solids or liquids around you? ______

2. Why do you think that is? ______

Matter

TARGETING SCIENCE YEAR 3 © PASCAL PRESS ISBN: 9781925726527

Look at the picture. Read the word.
Write the word in the sentence.

solid

A ____________________ has its own shape and does not flow.

liquid

A ____________________ flows but does not have its own shape.

gas

A ____________________ does not have its own shape and is invisible.

A **solid** has its own shape. A solid will not change unless you do something to it, like cut it or melt it.

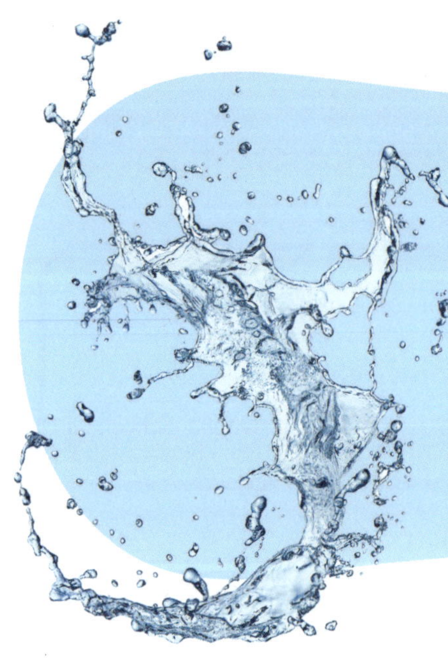

A **liquid** does not have its own shape. A liquid can flow, drip, and splash. It will take the shape of what it is in.

A **gas** is invisible. Air is made of gases. You cannot see air, but you can see what air does.

TARGETING SCIENCE YEAR 3 © PASCAL PRESS ISBN: 9781925726527

Skill:
Apply science vocabulary in context

Read the clue. Write the word.

1. A liquid can

f ___ ___ ___ .
(letter 4 is numbered 1)

2. It has its own shape.

___ ___ ___ ___ ___

3. It is made of gases.

___ ___ ___
(letter 1 is numbered 2; letter 3 is numbered 5)

4. It has three forms.

___ ___ ___ ___ ___
(letter 3 is numbered 3; letter 5 is numbered 4)

Write the numbered letters to solve the puzzle.

Science Puzzle

___ ___ ___ ___ ___ is a liquid.
1 2 3 4 5

TARGETING SCIENCE YEAR 3 © PASCAL PRESS ISBN: 9781925726527

Sorting Matter Chart

Skill:

Complete charts or graphs to categorise information

Write each word in the chart.

milk rock water air toy
ball juice balloon

solid	liquid	gas

Matter

TARGETING SCIENCE YEAR 3 © PASCAL PRESS ISBN: 9781925726527

Skill:

Conduct experiments and record results

What You Need

- 3 balloons of different colours
- water

What You Do

1. Fill one balloon halfway with water. Freeze overnight. The next day, take out the frozen balloon. This is your solid balloon. How does it feel? Write it on the next page.

2. Fill another balloon halfway with water. This is your liquid balloon. How does it feel? Write it on the next page.

3. Blow up another balloon. This is your gas balloon. How does it feel? Write it on the next page.

4. Toss each balloon. Write what happened on the next page.

Write what happened when you tossed the balloons.

What Happened?

Frozen

How did it feel? ______________________

What happened when I tossed it?

____

Liquid

How did it feel? ______________________

What happened when I tossed it?

____

Gas

How did it feel? ______________________

What happened when I tossed it?

____

TARGETING SCIENCE YEAR 3 © PASCAL PRESS ISBN: 9781925726527

Skill:

Complete charts or graphs to answer questions

Solids, liquids, and gases are kinds of matter. Complete the chart to show how they are alike or different.

Matter	Does it have its own shape?	Can you see it?	Write one example.
solid			
liquid			
gas			

Melting, Freezing, Mixing

Look at the picture. Read the word.
Write the word in the sentence.

melt

When you ________________ something, it goes from a solid to a liquid.

freeze

When you ________________ something, it goes from a liquid to a solid.

mixture

When you put different things together, you make a ________________.

TARGETING SCIENCE YEAR 3 © PASCAL PRESS ISBN: 9781925726527

Concept:
Matter can change when it is mixed, heated, or cooled.

Matter can change in different ways. It can change from a solid to a liquid. It can change from a liquid to a solid.

Heat can change matter. The heat from the stove changed this butter. The butter changed from solid to liquid. The butter **melted**.

Cold air can change matter. Cold air will **freeze** water. The water changed from a liquid to a solid. The water froze and became ice cubes.

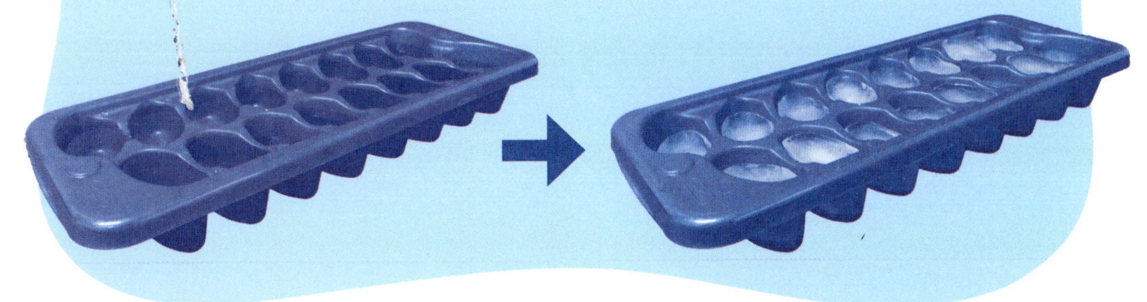

Another thing you can do with matter is mix it. You can put together different things and make a mixture.

a mixture of fruit

Things that make up a mixture do not change when you put them together.

This mixture has many different kinds of fruit. If you take a cherry out of the mixture, the cherry keeps the same shape that it had when it was in the mixture.

 TARGETING SCIENCE YEAR 3 © PASCAL PRESS ISBN: 9781925726527

Skill:
Apply science vocabulary in context

Read the clue. Write the word.

1. Solid to liquid

____ ____ ____ ____
          5   1

2. Liquid to solid

____ ____ ____ ____ ____ ____
     2

3. Two or more things mixed together

____ ____ ____ ____ ____ ____ ____
6   4   7

4. To become different

____ ____ ____ ____ ____ ____
          3

Use the numbered letters to solve the puzzle.

Changes in Matter

Changes in Matter

Skill:

Interpret information from graphic images

Look at the diagrams.

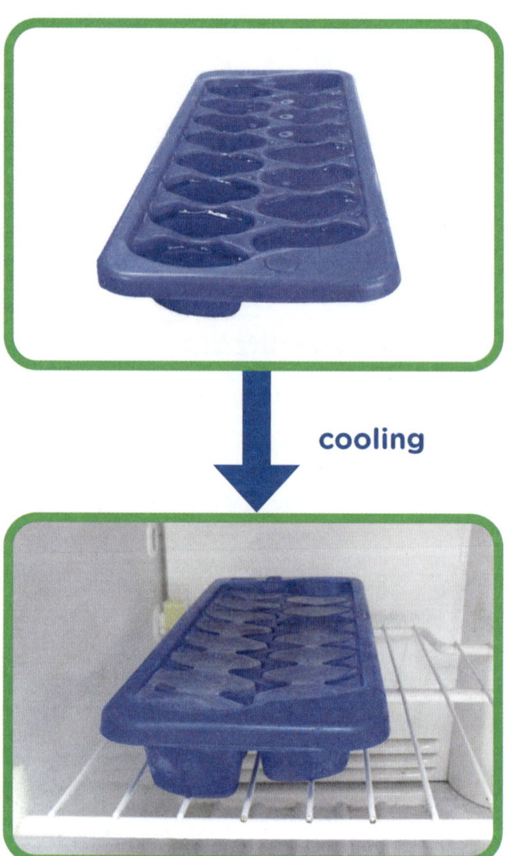

In what way did the butter change?

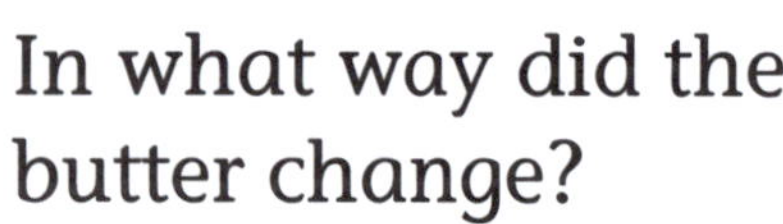

In what way did the water change?

Crayon Paperweight

What You Need

- crayons
- 2 smooth flat stones
- 2 paper plates

What You Do

This activity is best done on a very sunny day.

1. Peel the paper from the crayons. Make crayon shavings by using a pencil sharpener.
2. Place each stone on a paper plate. Sprinkle crayon shavings on top of each stone.
3. Set one crayon-covered stone in direct sunlight. Set the other crayon-covered stone in the shade.

4. What do you think will happen? Why?
5. After one hour, check the stones. What happened? Write or draw to tell about any changes.

Stone 1

What I did: ____________________

What happened: ____________________

____

What it looks like:

Stone 2

What I did: ____________________

What happened: ____________________

____

What it looks like:

TARGETING SCIENCE YEAR 3 © PASCAL PRESS ISBN: 9781925726527

Changes Chart

Skill:

Write sentences to demonstrate understanding of science concepts

Write a sentence about each picture.

Recycling Materials

Did you know that many of the things we use every day can be **recycled** and turned into something else? It takes a bit of practise to know what goes where, but once you get the hang of it, you can really make a difference to the amount of rubbish that ends up being buried underground!

The reason many of these things can be recycled is because they can be melted down into a liquid after they have been collected from the bins, and then they are turned into solids again.

Look at the process this paper goes through to become new paper:

When you throw a piece of paper or card into the correct bin, it is shredded up and mixed with chemicals to make it into a liquid again. Then it is turned into a brand new piece of paper or a box.

How good are you are sorting these rubbish items into the correct bins? Draw a line from the item to each bin:

TARGETING SCIENCE YEAR 3 © PASCAL PRESS ISBN: 9781925726527

First Nations Perspectives on Beeswax

https://clickv.ie/w/X0gx

Use this QR code to access a video on this topic.

First Nations Peoples loved the honey from stingless native bees. It was often called 'sugarbag' and traded between tribes. It was considered to be very valuable for many reasons.

They people saw the value of this honey as a food source, but knew it was also very good as a medicine for 'cleaning out the gut'.

The bees wax from inside the hive had special properties which made it particularly useful. It could be heated and cooled, so that it could be re-used over and over.

When a wick was placed in it, it burned as a candle.

Warm wax was used for sealing and filling holes in artifacts and weapons, and for the mouthparts of digeridoos.

Heated wax melts, and when it cools down, it sets hard. The wax could be warmed and shaped as many times as needed, until the mouthpiece was just right for the player's mouth.

From your reading, what are five ways First Nations Peoples use honey and wax from native bees?

1. ______________________

2. ______________________

3. ______________________

4. ______________________

5. ______________________

Answer Key

Page 3

zebra L, water NL, cut flowers OL, cheese NL, fire NL, bowl of fruit OL, pot plant L, feather NL, rocks NL

1. A zebra is a living thing. It can breathe, eat, respond, move, have babies, grow, and remove waste.
2. A stack of rocks is a non-living thing. It cannot breathe, eat, respond, move, have babies, grow, or remove waste.
3. A cut flower is a once-living thing. It can longer breathe, eat, respond, move, have babies, grow, and remove waste.

Page 4

Mammals - mother, milk, warm, fur

Fish - scales, blood, young, eggs

Birds - eggs, nest feathers

Reptiles - scales, cold, live, crocodiles

Insects - three, stages, metamorphosis

Amphibians - skin, eggs, completely, adults

Page 7

1 chick, 2 mammals, 3 milk, 4 nest, puzzle: hatches

Page 8

1 goose, 2 chicken, 3 The goose egg takes longer because the goose is bigger.

Page 9

Mother cat grows kittens inside her - Kittens are born - Kittens drink mother's milk - Kittens grow into cats

Mother hen lays eggs - Chicks grow inside the eggs -Chicks hatch - Chicks grow into adult chickens

Page 10

1 stamen, 2 germinate, 3 pollen, 4 reproduce, puzzle: seed

Page 12

1 seed coat, 2 germination, 3 leaves, roots, often flowers

Page 13

1 reproduce, 2 life cycle, 3 animals or the wind

Page 14

1, 3, 2, 5, 4

pollen, pollinates

Page 15

1b, 2d, 3a, 4c

Page 17

1. A seed is dropped onto the soil.
2. A plant begins to grow from the seed.
3. The plant develops roots, leaves and a flower.
4. The flower produces seeds and the seeds drop onto the soil. Then the cycle starts over.

Page 18

1 energy, 2 from food

Page 19

1 rabbit 9 turtle 120 or more, 2 rabbit 2 weeks turtle at birth

Page 20

1 false, 2 true

Page 21

(clockwise from top right) 2, 1, 3, 6, 5, 4, 7

Page 22

1 living, 2 baby bird, 3 food, 4 mammal, 5 offspring, 6 change, 7 hard shell, 8 moult, 9 cold-blooded, 10 larva

Page 25

1 sea turtle, 2 duckling, 3 giraffe, 3 human

Page 26

1 false, 2 true, 3 true

Page 27

chimp baby centre, orangutan baby right, squirrel monkey baby left

Page 29

2 insects are nutritious and readily available

Page 32

1 heat and pressure, 2 volcano, 3 weathering and lithification

Page 39

1 bone, 2 cast

Page 40

1 extinct, 2 fossil record

Page 41

1 false, 2 true, 3 false

Page 42

1 c, 2 a, 3 d, 4 b

Page 43

1 fossil, 2 landmass, 3 fern, 4 continents, 5 trilobite, 6 sediment, 7 decay, 8 extinct

Page 44

NSW Devonian fish, VIC Koolasuschus, ACT Trilobite, WA Gogo fish, SA Spriggina

TARGETING SCIENCE YEAR 3 © PASCAL PRESS ISBN: 9781925726527

Page 46

Answers may include the following ideas: the sea animal dies and is quickly buried by sediment; more layers build up and a fossil is made; the continents move together and push up the ocean floor; the rock layer with the fossil in it is pushed up and becomes a tall mountain.

Page 49

1 surface, 2 erosion, 3 weathering, 4 changes, puzzle: forces

Page 50

Cut down trees, dug up big rocks, made the land flat (the others may also apply depending on what the land was like before)

Page 51

Shaking the jar is similar to weathering, as the rocks are being rubbed against each other and against the jar.

Page 52

3 some of the particles moved, 5 the soil became wet and some particles washed away, 6 you could build walls around it to hold the particles in place

Page 53

people, erosion, weathering

page 54

1 false, 2 true, 3 true, 4 false

Page 55

1 volcanic rocks, 2 earthquakes, 3 Earth was made of different layers of different densities and some of the interior was liquid

Page 56

battery MM, sun N, lightning N, bushfire N, air-conditioner MM, oven MM, match MM, firefly N, torch MM, lightbulb MM, chemical reaction N or MM, heater MM, stove MM, star N

Page 57

1 warm to cool, 2 move faster

Page 58

1 they are moving at the same speed, 2 conduction

Page 59

1 true, 2 true, 3 false

Page 60

top left radiation, top right convection currents, bottom conduction

Page 61

1 particles of matter, 2 how cold it is, 3 radiation, 4 breezes, 5 conductor, 6 the sun, 7 currents, 8 different

Page 63

1 towards the cooler blue-coloured water, 2 heat energy flows from warm objects to cooler ones, 3 convection currents, 4 they started to mix and turn purple, 5 they all have the same temperature

Page 64

Answers will vary, but should show an understanding that heat energy makes molecules spread apart and move faster, and that convection currents start to form when heated liquid rises.

Page 66

It is likely that the ice that melted the most was wrapped in tissue, the one in paper melted a bit less, and the one in foil the least. Foil is the best insulator. Paper is also a good insulator until it becomes wet.

Page 68

Answers will vary. Solids hold their shape while liquids flow and take the shape of their container.

1 it is easier to find solids, 2 solids are more plentiful

Page 71

1 flow, 2 solid, 3 air, 4 water, puzzle: water

Page 72

solid: rock, toy, ball; **liquid:** milk, water, juice; **gas:** air, the inside of an inflated balloon

Page 74

frozen: the water expanded when it turned to ice, it felt cold and solid, it fell heavily when tossed

liquid: the balloon felt squishy and floppy, it bounced a bit when tossed then it broke

gas: the balloon felt firm but light, it floated a but when tossed then drifted and bounced

Page 75

solid: yes, yes, rock; **liquid:** no, yes, milk; **gas:** no, no, air

Page 79

1 melt, 2 freeze, 3 mixture, 4 change, puzzle: trail mix

Page 80

The butter melted or went from a solid to a liquid, The water froze or went from a liquid to a solid.

Answer Key

Page 82

stone 1: placed in sunlight, crayon shavings melted onto the rock and made patches of colour

stone 2: placed in shade, the crayon shavings didn't melt and the rock's appearance didn't change

Page 83

Answers will vary, but should reflect any changes in the state of matter that may arise from: 1 melting, 2 mixing, 3 freezing

Page 84

Images left to right: box-paper, can-metal, bottle-hard plastic and glass, newspaper-paper, pot-metal, wrapping paper-paper

Page 85

1 trade, 2 food, 3 medicine, 4 candles, 5 sealing and filling holes, also mouthparts of didgeridoos